Nuclear Medicine Technology

Eleanor Mantel • Janet S. Reddin
Gang Cheng • Abass Alavi

Nuclear Medicine Technology

Review Questions for the Board Examinations

Sixth Edition

Eleanor Mantel
Nuclear Medicine and Radiology
University of Pennsylvania
Philadelphia, PA, USA

Janet S. Reddin
Nuclear Medicine and Radiology
University of Pennsylvania
Philadelphia, PA, USA

Gang Cheng
Department of Radiology
Philadelphia VA Medical Center
Philadelphia, PA, USA

Abass Alavi
Nuclear Medicine and Radiology
University of Pennsylvania
Philadelphia, PA, USA

ISBN 978-3-031-26719-2 ISBN 978-3-031-26720-8 (eBook)
https://doi.org/10.1007/978-3-031-26720-8

This Springer imprint is published by the registered company Springer Nature Switzerland AG
The registered company address is: Gewerbestrasse 11, 6330 Cham, Switzerland

Contents

Introduction

1

Since Marie Curie named the mysterious rays "radioactivity" in 1897, nuclear medicine has grown exponentially, paving the way for nuclear medicine as we know it today, with the growth of positron-emission tomography, radionuclide therapies, and other molecular imaging. Taking and passing the exam offered by the Nuclear Medicine Technology Certification Board (NMTCB), or that offered by the American Registry of Radiologic Technologists (ARRT), is the final step in reaching the status of nuclear medicine technologist. Both of these exams are challenging, and the breadth of knowledge that they cover means that a thorough review is in order before attempting either exam.

This book was created to assist in the preparation for those exams. In addition to recalling information, the questions require application of information and analysis of situations. This edition provides detailed explanations for answers to the questions in each chapter and offers a new chapter on positron-emission tomography. (Since 1978, the NMTCB has been offering a specialty exam in this modality, allowing technologists to continue their education and augment their career qualifications.)

Both the ARRT and the NMTCB use computer testing for these exams. Because of this, questions cannot be skipped. A good strategy is to try to get through all of the questions and marking or flagging each one that was a guess. In the event that another question provides you with information that changes your

E. Mantel et al., *Nuclear Medicine Technology*,
https://doi.org/10.1007/978-3-031-26720-8_1

guess later, you will be able to return to that question and change your answer. If you have not marked it, you may not have time to find it.

It is well worth the time (even if one is extremely comfortable using computers) to take time for the tutorials offered before the exam timer begins. This will familiarize you with the location of buttons and functions on the screens and may save you a few minutes of navigating during the actual exam especially while reviewing.

As was the case with paper exams, careful reading of the question cannot be overstressed. Consider, for example, the difference in being asked to state what distance must be maintained to reduce exposure to a radioactive source by 75% as opposed to being reduced to 75% of the original. It is also extremely important to ask oneself whether the answer makes sense at the end of a calculation. For instance, if the question is about the amount of radioactivity present at some time prior to an assay and if the calculation does not result in an amount greater than the assayed amount, a recalculation is in order. When completely stumped by a question, try to rule out a few of the answers offered, thereby increasing your chances of a correct guess. As was the case with paper exams, there is no penalty for a wrong guess, and so it is always better to give any answer than to give none. Pacing to get through all the questions is therefore important.

All the best in reviewing, testing, and performing as a certified or registered nuclear medicine technologist!

Radioactivity, Radiopharmacy, and Quality Assurance

2

1. How does ^{201}Tl decay?

   (a) By positron emission
   (b) By beta emission
   (c) By internal conversion
   (d) By electron capture

2. What is the role of the stannous ion in the preparation of pharmaceuticals labeled with ^{99m}Tc?

   (a) To increase the valence state from +4 to +7
   (b) To reduce the amount of $A1^{+3}$ present
   (c) To reduce the valence state of ^{99m}Tc
   (d) To reduce the radiation dose

3. If an assay of a vial containing ^{131}I shows 50 mCi present on May 2, approximately what will the assay show on May 18?

   (a) 25 mCi
   (b) 12.5 mCi
   (c) 40 mCi
   (d) 6 mCi

4. If a bone scan has been ordered on a 10-year-old weighing 32 kg, and the physician prescribes 0.2 mCi/kg be given, how many mCi should be administered?

   (a) 2.0 mCi
   (b) 5.1 mCi

E. Mantel et al., *Nuclear Medicine Technology*,
https://doi.org/10.1007/978-3-031-26720-8_2

(c) 6.4 mCi
(d) 3.1 mCi

5. If the biologic half-life of an isotope is 6 h and the physical half-life is 12 h, what is the effective half-life?

   (a) 6 h
   (b) 12 h
   (c) 2 h
   (d) 4 h

6. Which of the following is used to abbreviate physical half-life?

   (a) T_p
   (b) T/2
   (c) T_2
   (d) P½

7. The physical half-life of a radionuclide is the time it takes:

   (a) For half of the substance to leave the body
   (b) For the nuclide to decay to one-half of the original activity
   (c) For the kit to become half expired
   (d) For half of the substance to be metabolized

8. If a kit has 310 mCi of activity present at 8:00 a.m., what will the vial assay show in 4 h and 10 min if the decay factor is 0.618?

   (a) 175 mCi
   (b) 192 mCi
   (c) 501 mCi
   (d) 155 mCi

9. A vial containing ^{99m}Tc is assayed at 9:00 a.m. and contains 255 mCi. To calculate the remaining activity at 3:00 p.m., what decay factor would be used?

   (a) 2.0
   (b) 0.8
   (c) 0.5
   (d) 0.2

10. A vial of technetium eluate contains 50 mCi/mL. If 4 mL is withdrawn and added to a diphosphonate kit containing 16 mL of solution, what volume would then need to be withdrawn to prepare a 20 mCi dose at that moment?

    (a) 1.0
    (b) 1.5
    (c) 2.0
    (d) 2.5

11. If a preparation of ^{99m}Tc mertiatide has 60 mCi of activity present at 8:30 a.m., how many mCi will be present at 9:00 a.m. (DF = 0.944)?

    (a) 63.6
    (b) 56.6
    (c) 59.6
    (d) 53.6

12. Which of the following is boiled during preparation?

    (a) MAA
    (b) Sulfur colloid
    (c) Albumin colloid
    (d) Diphosphonates

13. The presence of 12 μg Al^{+3} in 1 mL of ^{99m}Tc eluate is:

    (a) An example of radionuclidic impurity
    (b) An example of chemical impurity
    (c) An example of radiochemical impurity
    (d) Acceptable since it is less than 15 μg/mL

14. Which body decides on the acceptable levels of radionuclidic impurity?

    (a) DEP
    (b) NRC
    (c) FDA
    (d) AEC

15. Which of the following is an example of radionuclidic impurity?

    (a) Presence of free ^{99m}Tc in a preparation of ^{99m}Tc sulfur colloid
    (b) Presence of ^{99}Mo in ^{99m}Tc eluate
    (c) Presence of aluminum ions in ^{99m}Tc eluate
    (d) Presence of pyrogens in eluate

16. What is the maximum amount of aluminum ions (Al^{+3}) allowed in 1 mL of ^{99m}Tc eluate according to the USP?

    (a) None is allowed
    (b) 5 µg
    (c) 10 µg
    (d) 15 µg

17. What is indicated by the front of an instant thin-layer chromatography (ITLC) strip?

    (a) Radionuclidic impurity
    (b) Particles of incorrect size
    (c) Pyrogens
    (d) This depends on the solvent and strip used

18. If a kit contains 140 mCi of ^{99m}Tc in 23 mL, how much volume must be withdrawn to obtain a dose of 5 mCi?

    (a) 0.8 mL
    (b) 30 mL
    (c) 1.2 mL
    (d) 0.6 mL

19. If a kit contains 140 mCi of ^{99m}Tc in 23 mL at 9:00 a.m., how much volume must be withdrawn to obtain a dose of 5 mCi at 3:00 p.m.?

    (a) 0.8 mL
    (b) 1.6 mL
    (c) 2.4 mL
    (d) 0.6 mL

20. An MAA kit contains 40 mCi of ^{99m}Tc in 5 mL at 8:00 a.m. What would be the best volume to be withdrawn for a 4 mCi dose at 10:00 a.m. if a perfusion lung scan is planned (DF = 0.794)?

    (a) 0.63 mL
    (b) 1.54 mL
    (c) 2.2 mL
    (d) 0.25 mL

21. What is the most likely size of an MAA particle if correctly prepared?

    (a) 0–100 mm
    (b) 10–30 µm
    (c) 10–30 mm
    (d) 0–250 µm

22. ^{99m}Tc MAA has a biologic half-life of 2–4 h. What will the effective half-life be?

    (a) 1.5–3.0 h
    (b) 2.0–4.0 h
    (c) 0.5–1.0 h
    (d) 1.5–2.4 h

23. Which radiopharmaceutical is made with ^{99m}Tc without a reducing agent?

    (a) MAG3
    (b) MAA
    (c) Sulfur colloid
    (d) Sestamibi

24. Which of the following is an example of radiochemical impurity?

    (a) Presence of free ^{99m}Tc in a preparation of ^{99m}Tc sulfur colloid
    (b) Presence of ^{99}Mo in ^{99m}Tc eluate
    (c) Presence of aluminum ions in ^{99m}Tc eluate
    (d) Presence of pyrogens in eluate

25. Which of the following can be said regarding effective half-life?

    (a) It is always longer than the physical half-life.
    (b) It is always shorter than both the physical and the biologic half-life.
    (c) It is always shorter than physical half-life but longer than the biologic half-life.
    (d) It is always longer than the biologic half-life but shorter than the physical half-life.

26. The purpose of adding EDTA to sulfur colloid when labeling with ^{99m}Tc is:

    (a) To prevent aggregation of sulfur colloid
    (b) To bind excess Al^{+3}
    (c) To prevent loss of the radiolabel
    (d) (a) and (b) only
    (e) (b) and (c) only

27. A diphosphonate kit should generally be used within how many hours after preparation?

    (a) 2 h
    (b) 12 h
    (c) 4–6 h
    (d) 24 h

28. What is the usual particle size of sulfur colloid?

    (a) 0.03–0.1 μm
    (b) 0.3–1.0 μm
    (c) 2.0–10 μm
    (d) 4.0–15 μm

29. Which radiopharmaceutical, when correctly prepared, will have the smallest particle size?

    (a) ^{99m}Tc sulfur colloid
    (b) ^{99m}Tc albumin colloid
    (c) ^{99m}Tc human serum albumin
    (d) ^{99m}Tc macroaggregated albumin

30. The advantages of albumin colloid over sulfur colloid include which of the following?

    (a) Does not require heating
    (b) Less expensive
    (c) Smaller dose can be administered

31. Following injection of ^{99m}Tc MAA for a perfusion lung scan, activity is seen in the kidneys and brain. This is indicative of:

    (a) Right-to-left cardiac shunt
    (b) Renal failure
    (c) Congestive heart failure
    (d) Incorrect particle size

32. At 7:00 a.m., a technologist prepares a dose of ^{99m}Tc MDP for injection at 10:00 a.m. that day. The desired dose is 22 mCi and no precalibration factors are available. The 3 h decay factor for the isotope is 0.707. What amount of activity should the technologist draw up into the syringe at 7:00 a.m.?

    (a) 15.6 mCi
    (b) 27.07 mCi
    (c) 29.5 mCi
    (d) 31.1 mCi

33. What can be said regarding precalibration factors?

    (a) It is not necessary for problem-solving if the decay factor is available
    (b) It is always <1.0
    (c) It is always >1.0
    (d) Both (a) and (c)

34. What method is used to calculate pediatric dose?

    (a) According to weight
    (b) Clark's formula
    (c) According to body surface area
    (d) Using Talbot's nomogram
    (e) All of the above

35. If the recommended volume for a MAG3 kit ranges from 4 to 10 mL, and the ^{99m}Tc eluate that will be used contains 820 mCi in 10 mL, and 41 mCi will be used, what is the minimum amount of diluent that should be added?

    (a) 0.5 mL
    (b) 1 mL
    (c) 3.5 mL
    (d) 9.5 mL

36. If a 20 mCi dose of ^{99m}Tc HDP is needed at 9:00 a.m., how much activity should the syringe contain if the technologist prepares it at 7:00 a.m.? You may use the table of precalibration factors (Table 2.1) to determine the answer.

    (a) 15.9 mCi
    (b) 21.259 mCi
    (c) 25.18 mCi
    (d) 26.7 mCi

37. Using Table 2.2, determine the decay factor for ^{99m}Tc at 7 h.

    (a) 1.337
    (b) 0.445
    (c) 0.432
    (d) 0.551

38. On a Monday morning at 6:00 a.m. a technologist is preparing a ^{99m}Tc ECD kit that is to be used for SPECT brain scan injections at 8:00 a.m., 9:00 a.m., and 10:00 a.m. Each patient should receive 10 mCi. What is the minimum activity that should be added to the kit during preparation? Use Table 2.1 if necessary.

    (a) 42.6 mCi
    (b) 30.0 mCi
    (c) 44.5 mCi
    (d) 52.0 mCi

Table 2.1 Precalibration factors for ^{99m}Tc (assuming $T_{1/2} = 6.0$ h)

	0:00	00:15	00:30	00:45
1:0	1.122	1.156	1.189	1.224
2:0	1.259	1.297	1.335	1.374
3:0	1.414	1.456	1.499	1.543
4:0	1.587	1.634	1.681	1.730

Table 2.2 Decay factors for ^{99m}Tc (assuming $T_{1/2} = 6.0$ h)

	0:00	00:15	00:30	00:45
1.0	0.891	0.866	0.841	0.817
2.0	0.794	0.771	0.749	0.728
3.0	0.707	0.687	0.667	0.648
4.0	0.630	0.612	0.595	0.578

39. A chromatography strip is used to test a kit for radiochemical impurity and is counted in a well counter. Part A contains ^{99m}Tc pertechnetate, and Part B contains bound ^{99m}Tc in the desired form. If the results show 258,000 cpm in Part B and 55,000 cpm in Part A, can this kit be used for injection into patients?

    (a) Yes
    (b) No

40. What is the approximate radiochemical purity for the kit described in Question 39?

    (a) 21%
    (b) 79%
    (c) 18%
    (d) 82%

41. What is the approximate radiochemical impurity of the kit described in Question 39?

    (a) 21%
    (b) 79%
    (c) 18%
    (d) 82%

42. A vial of ^{99m}Tc eluate is tested for ^{99}Mo breakthrough, and the amount of breakthrough is 25 μCi in 775 mCi at 6:00 a.m. Following the preparation of all kits to be used that day, 450 mCi of ^{99m}Tc is left. That night, a technologist is asked to perform a scrotal scan at 11:00 p.m. Must the generator be eluted again?

    (a) Yes, because the amount of eluate will have decayed to below the amount needed for a patient dose.
    (b) Yes, because the molybdenum breakthrough will now exceed the limit allowed by the NRC.
    (c) No.

43. A ^{99m}Tc MDP bone scan dose was prepared at 7:00 a.m. and contained 32 mCi/2 mL. At 9:00 a.m., when the patient arrives, the technologist realizes that the patient's age was overlooked (13 years). The technologist would now like to adjust the dose to 11 mCi. Given a 2-h decay factor of 0.794, what volume should be discarded so that the correct dose remains in the syringe?

    (a) 0.65 mL
    (b) 0.87 mL
    (c) 1.13 mL
    (d) 1.5 mL

44. A dose of ^{99m}Tc DMSA is prepared and calibrated to contain 5.0 mCi at 8:00 a.m. The patient arrives late at 10:00 a.m. Without using any tables of decay factors, determine what activity will remain in the dose at that time.

    (a) 3.40 mCi
    (b) 3.54 mCi
    (c) 3.62 mCi
    (d) 3.97 mCi

45. An MAA kit contains 950,000 particles per mL. The activity in the kit is 50 mCi of ^{99m}Tc in 5 mL. If a 4 mCi dose is drawn up, how many particles will be in the dose?

    (a) 76,000
    (b) 380,000
    (c) 410,000
    (d) 450,000

46. What will happen to the dose in Question 45 if it sits for 1 h?

    (a) The number of particles per mCi will increase.
    (b) The number of particles per mCi will decrease.
    (c) The number of particles per mCi will remain unchanged.

47. A volume of 5 mL containing 40 mCi of ^{99m}Tc is added to an MAA kit with an average of 3,000,000 particles. What volume of the reconstituted kit should be withdrawn to prepare a dose for a patient with severe pulmonary hypertension?

    (a) 0.25 mL
    (b) 0.40 mL
    (c) 0.45 mL
    (d) 0.50 mL

48. To reduce the possibility of pyrogenic reactions, all kits should be prepared using saline that contains bacteriostatic preservatives.

    (a) True
    (b) False

49. While performing a GI bleeding study with labeled red blood cells, a technologist notices gastric activity that he or she suspects is the result of free pertechnetate. What could be done to support this suspicion?

    (a) Reimage the patient in the erect position.
    (b) Narrow the window around the photopeak.
    (c) Image the thyroid.
    (d) Have the patient drink two glasses of water and empty his or her bladder.

50. Convert 23 mCi to SI units.

    (a) 851 MBq
    (b) 851 kBq
    (c) 851 GBq
    (d) 0.62 MBq

51. If excessive aluminum is present in ^{99m}Tc eluate, which of the following would be expected on a bone scan?

    (a) Lung uptake
    (b) Liver uptake
    (c) Thyroid uptake
    (d) Gastric uptake

52. Radiochemical impurities often result from:

    (a) Introduction of water into the kit
    (b) Introduction of oxygen into the kit
    (c) Introduction of nitrogen into the kit
    (d) (a) and (b)

53. It is proper technique to clean the septum of a kit reaction vial and inject an amount of air equal to the volume being withdrawn when preparing a unit dose.

    (a) True
    (b) False

54. 15 rem is equal to:

    (a) 150 mSv
    (b) 15 gray
    (c) 15 Sv
    (d) 150 MBq

55. What is the purpose of adding hetastarch to a blood sample drawn for the purpose of leukocyte labeling?

    (a) To act as an anticoagulant
    (b) To hasten the settling of erythrocytes
    (c) To separate platelets from leukocytes
    (d) To improve labeling efficiency

56. Following reconstitution of a kit with ^{99m}Tc pertechnetate, a technologist should ensure that all of the following are present on the vial except:

    (a) Date and time of preparation
    (b) Lot number
    (c) Concentration and volume
    (d) Patient name or identification number

57. If the proper centrifuge speed is not used during separation of cell types for leukocyte labeling with ^{111}In oxine, what may happen?

    (a) Platelets may be inadvertently labeled.
    (b) ^{111}In oxine will not tag WBCs.
    (c) Red blood cells may become damaged.
    (d) White blood cells may become damaged.

58. In most reconstituted radiopharmaceutical kits, in what form is ^{99m}Tc present?

    (a) Free pertechnetate
    (b) Bound technetium
    (c) Reduced, hydrolyzed technetium
    (d) All of the above

59. For determination of plasma volume, 10 μCi of human serum albumin in 2.5 mL is added to 500 mL of water. What is the concentration of the resulting solution?

    (a) 0.019 μCi/mL
    (b) 0.020 μCi/mL
    (c) 50.00 μCi/mL
    (d) 50.20 μCi/mL

60. A ^{99}Mo/^{99m}Tc generator exists in _____ equilibrium, and the parent isotope has a _____ physical half-life than the daughter isotope.

    (a) Secular, longer
    (b) Secular, shorter
    (c) Transient, longer
    (d) Transient, shorter

61. If 630 mCi of ^{99m}Tc is eluted from a ^{99}Mo/^{99m}Tc generator, what is the NRC limit of total ^{99}Mo activity that may be present?

    (a) 0.15 μCi
    (b) 94.5 μCi
    (c) 42 mCi
    (d) 94.5 μCi/mL
    (e) 42 μCi

3 Radiation Safety

1. Which of the following bodies regulates transportation of radiopharmaceuticals?
   (a) NRC
   (b) DOT
   (c) TJC
   (d) FDA

2. Which of the following bodies regulates the use of investigational pharmaceuticals?
   (a) NRC
   (b) DOT
   (c) IRB
   (d) FDA

3. In the event of a spill of ^{99m}Tc onto clothing, one should immediately:
   (a) Enter a shower fully clothed.
   (b) Remove and store the clothing until they decay to background.
   (c) Wash the clothing in hot water and then survey them to determine the remaining activity.
   (d) Remove and destroy the clothing.

E. Mantel et al., *Nuclear Medicine Technology*,
https://doi.org/10.1007/978-3-031-26720-8_3

4. If a radiopharmaceutical is spilled on the floor, the first priority is to:

   (a) Contact the Radiation Safety Officer.
   (b) Pour a chelating solution over the area of the spill.
   (c) Cover the area with absorbent paper and restrict access around it.
   (d) Call the housekeeping department to arrange for cleaning.

5. The inverse square law, in words, says:

   (a) If you double the distance from the source of activity, you reduce exposure to 25% of the original intensity.
   (b) If you halve the distance from the source of activity, you decrease exposure to 25% of the original intensity.
   (c) If you halve the distance from the source of activity, you decrease exposure to one-fourth of the original intensity.

6. What is the best way to decrease the radioactive dose to visitors if a patient is surveyed to emit 3 mrem/h at bedside?

   (a) Have the patient wear a lead apron.
   (b) Keep the patient well hydrated and encourage frequent voiding.
   (c) Have the visitor sit or stand as far as possible from bedside.
   (d) Have the visitor wear lead shielding.

7. Which of the following isotopes would be effectively shielded by a plastic syringe?

   (a) ^{67}Ga
   (b) ^{89}Sr
   (c) ^{99m}Tc
   (d) ^{81m}Kr
   (e) ^{133}Xe

8. What is the NRC annual dose limit allowed to the lens of the eye?

   (a) 1.5 mrem
   (b) 15 rem
   (c) 50 rem
   (d) 5 rem

9. Which of the following should be used when administering an intravenous pharmaceutical to a patient?

   (a) Lead syringe shield
   (b) Leaded eyeglasses
   (c) Gloves
   (d) (a) and (b) only

10. Which of the following is the most effective means of measuring low levels of removable radiation?

    (a) By performing an area survey
    (b) By performing a wipe test
    (c) With a pocket dosimeter
    (d) With a TLD

11. What is the dose rate limit at the package surface for a shipment of radioactive material bearing a Yellow-III label?

    (a) 200 mrem/h
    (b) 50 mrem/h
    (c) 200 rads
    (d) 200 mrem

12. Which of the following measures absorbed doses?

    (a) mCi
    (b) Becquerel
    (c) Gray

13. If the dose rate at 3 m from a radioactive source is 100 mrem/h, what will the dose rate be at 6 m?

    (a) 25 mrem/h
    (b) 50 mrem/h
    (c) 75 mrem/h
    (d) 12.5 mrem/h

14. The philosophy of the ALARA program is to keep the radiation dose:

    (a) As low as recently authorized
    (b) As long as reasonably attained
    (c) As long as reasonably acceptable
    (d) As low as reasonably achievable

15. All of the following are critical factors in keeping radiation exposure to a minimum except:

    (a) Time spent near the radioactive source
    (b) Geometry of the container holding the source of radiation
    (c) Distance from the source of radiation
    (d) Shielding of the radioactive source

16. Gaseous radiopharmaceuticals may only be used in rooms that:

    (a) Have at least one window
    (b) Contain an oxygen supply
    (c) Are at a positive pressure compared to surrounding rooms
    (d) Are at a negative pressure compared to surrounding rooms

17. If the exposure rate at 4 m from a radioactive source is 5 mR/h, what will the exposure rate be at 3 m?

    (a) 2.8 mR/h
    (b) 6.5 mR/h
    (c) 7.4 mR/h
    (d) 8.9 mR/h

18. A spill of ^{99m}Tc increases the exposure rate in a room from 1.7 to 3.15 mR/h. The room is posted with a sign reading "Caution-Radioactive Materials." What would be the ideal solution?

    (a) Change the sign to one reading "Caution-Radiation Area."
    (b) Call the NRC.
    (c) Decontaminate the floor with water and cleanser.
    (d) Place absorbent paper over the spill and close the room until the activity has decayed.

19. A technologist has 500 mrem registered on his or her ring badge in 1 month. What should be done to decrease exposure in the future?

    (a) Use lead pigs and syringe shields when preparing radiopharmaceuticals that emit gamma rays.

(b) Have another technologist elute the generator.
(c) Wear a lead apron.

20. OSHA requires that personnel exposure records be provided to employees:

(a) Monthly
(b) Quarterly
(c) Annually
(d) Biannually

21. A room containing a ^{57}Co sheet source is posted with a sign reading "Caution-Radioactive Materials." The exposure rate measured 30 cm from the source is 5.2 mrem/h. Which is the best solution?

(a) Change the sign to one reading "Caution-Radiation Area."
(b) Store the source in a lead-shielded container.
(c) Monitor the length of time a technologist can work near the source.

22. A technologist discovers that a patient in the room next to a radioiodine therapy will receive 2.5 mrem/h when lying in his or her bed, which is against the shared wall. Which is the best solution?

(a) Move the bed to the other side of the room.
(b) Discharge the therapy patient.
(c) Discharge the non-therapy patient.
(d) Calculate how long visitors to the non-therapy patient can stay.
(e) Calculate how long the patient may stay in bed each hour.

23. What is the dose rate limit at the surface of a package bearing a DOT Class I White label?

(a) 0.5 mrem/h
(b) 2 mrem/h
(c) 50 mrem/h
(d) 100 mrem/h
(e) 200 mrem/h

24. A type A package bears a DOT Class II Yellow Radioactive label, has a transport index of 0.8, and contains 10 mCi of ^{111}In oxine. What is the exposure rate at 1 m from the package?

    (a) 0.4 mrem/h
    (b) 0.8 mrem
    (c) 0.8 mrem/h
    (d) 7 mrem/h

25. The doorway to the nuclear medicine reception area should be posted with:

    (a) Caution: Radiation Area
    (b) Caution: High Radiation Area
    (c) Grave Danger: Very High Radiation Area
    (d) Caution: Radioactive Materials
    (e) None of the above

26. Which of the following steps would not decrease a technologist's chances of internal exposure to radiation?

    (a) Wearing gloves during injection of radiopharmaceuticals
    (b) Using tongs to transfer a vial from a lead shield to a dose calibrator
    (c) Working under a fume hood when working with volatile liquids and radioactive gases
    (d) Refraining from smoking and eating in the "hot lab."

27. Tools for measuring personnel exposure to radiation include all of the following except:

    (a) Thermoluminescent dosimeter
    (b) Pocket ionization chamber
    (c) Film badge
    (d) Geiger-Müller counter

28. Which of the following must be done during disposal of a carton in which a shipment of ^{131}I was received?

    (a) RSO must be notified.
    (b) Carton must be stored for ten half-lives before disposal.
    (c) Radioactive labels must be removed or obliterated.
    (d) Carton must be discarded with biohazardous waste.

29. If the lead HVL for ^{99m}Tc is 2.6 mm and a lead shield containing ^{99m}Tc eluate is 13 mm thick, what will the exposure rate be from the shielded vial if the unshielded vial had a rate of 100 mR/h?

    (a) 1.6 mR/h
    (b) 3.1 mR/h
    (c) 6.3 mR/h
    (d) 12.5 mR/h

30. Reports of area surveys must include all of the following except:

    (a) A diagram of the areas surveyed
    (b) Equipment that was used to perform the survey
    (c) Date performed
    (d) Initials of the person who performed the survey
    (e) List of isotopes used in the area

31. A technologist is working in a hot lab where the exposure rate is 20 mrem/h. What sign should be on the door?

    (a) Caution: Radioactive Materials
    (b) Caution: Radiation Area
    (c) Caution: High Radiation Area
    (d) Grave Danger: Very High Radiation Area
    (e) None of the above

32. If a technologist sits 2 ft away from a generator and the dose rate at his or her chair is 15 mrem/h, what will the dose rate be if he or she moves his or her chair 4 ft from the generator?

    (a) 2.5 mrem/h
    (b) 3.75 mrem/h
    (c) 7.5 mrem/h
    (d) 60 mrem/h

33. A medical event must be:

    (a) Reported within a week to the RSO
    (b) Reported to the NRC in writing within 15 days after discovery
    (c) Reported to the referring physician within 24 h

(d) Recorded and the records kept for 5 years
(e) All of the above except (a)

34. Which of the following is a medical event?

(a) When a dose of 400 μCi of ^{99m}Tc MAA is given rather than 4 mCi
(b) When the correct dose of ^{111}In oxine for cisternogram is injected intravenously
(c) When a patient who should have been injected with 20 mCi of ^{99m}Tc HDP for a bone scan receives a capsule containing 250 μCi of ^{123}I intended for a thyroid patient.
(d) When a patient who should have received 50 mCi of ^{131}I receives 35 mCi instead
(e) None of the above

35. Which of the following must be kept for 5 years?

(a) Records of medical events
(b) Records of dose calibrator linearity
(c) Records of doses assayed before administration
(d) Area survey records
(e) Records of instructions to breastfeeding women who received radiopharmaceuticals

36. Records of ^{99}Mo breakthrough in ^{99m}Tc eluate must be kept for:

(a) 1 year
(b) 3 years
(c) 5 years
(d) 10 years
(e) Until license expires

37. A pregnant technologist receives 350 mrem during her pregnancy according to a film badge worn at waist level. Has the NRC dose limit for the fetus been exceeded?

(a) Yes
(b) No

38. What is the best choice for disposal of a vial containing 27 mCi of ^{99m}Tc in 2 mL of liquid?

    (a) Transfer to an authorized recipient
    (b) Incinerate
    (c) Bury
    (d) Store until decayed to background
    (e) Release to the atmosphere through evaporation

39. Which of the following materials is sufficient for shielding a therapeutic dose of ^{90}Y ibritumomab tiuxetan?

    (a) Lucite
    (b) Tungsten
    (c) Lead
    (d) None of the above

40. Any patient who is treated with ^{90}Y ibritumomab tiuxetan must be hospitalized for at least 24 h.

    (a) True
    (b) False

4 Instrumentation and Quality Assurance

1. The abbreviation COR stands for:
   (a) Circle of rotation
   (b) Center of region
   (c) Correction of rotation
   (d) Center of rotation

2. How often should the COR check be performed?
   (a) At least weekly
   (b) Before each patient
   (c) Every 6 months
   (d) At least quarterly

3. What is the minimum number of counts that should be obtained in a uniformity correction flood for a SPECT camera?
   (a) 10 K
   (b) 10 million
   (c) 41 million
   (d) 60 million

4. Which of the following collimators will magnify an image?
   (a) Flat field
   (b) Diverging
   (c) Converging
   (d) High resolution

E. Mantel et al., *Nuclear Medicine Technology*,
https://doi.org/10.1007/978-3-031-26720-8_4

5. Which of the following must be performed every 6 months at minimum?
   (a) COR
   (b) Sealed source leak test
   (c) Linearity of the dose calibrator
   (d) Area survey of hot lab

6. A dose calibrator must undergo repair or replacement if accuracy or constancy errors are:
   (a) Present
   (b) Greater than 5%
   (c) Greater than 10%
   (d) Greater than 20%

7. Accuracy of a dose calibrator must be checked upon installation and then:
   (a) Monthly
   (b) Quarterly
   (c) Semiannually
   (d) Annually

8. Linearity of a dose calibrator must be checked upon installation and then:
   (a) Monthly
   (b) Quarterly
   (c) Semiannually
   (d) Annually

9. Constancy of a dose calibrator must be tested:
   (a) Daily
   (b) Quarterly
   (c) Weekly
   (d) Every 6 months

10. A technologist wishes to evaluate the intrinsic uniformity of a camera used for planar imaging and is preparing a dose of ^{99m}Tc. How much activity is sufficient for the image?
    (a) 2 mCi
    (b) 200 μCi

(c) 20 mCi
(d) 10 mCi

11. The advantages of using a ^{57}Co sheet source over a fluid-filled flood source with ^{99m}Tc added include:

    (a) It does not have to be prepared each day
    (b) Lower cost
    (c) Shorter half-life
    (d) Higher energy

12. Geometric dependence of a dose calibrator must be checked:

    (a) At installation
    (b) Daily
    (c) Quarterly
    (d) Annually

13. What happens to image resolution as the distance between a patient and a parallel-hole collimator decreases?

    (a) Improves
    (b) Worsens
    (c) Does not change

14. What happens to the image size as the distance between a patient and a parallel-hole collimator decreases?

    (a) Increases
    (b) Decreases
    (c) Does not change

15. What component of a gamma camera emits light when it has absorbed a photon?

    (a) Photomultiplier tube
    (b) Pulse height analyzer
    (c) Scintillation crystal
    (d) Collimator

16. Most of the photons emitted from the radiopharmaceutical which has been administered to a patient:

    (a) Do not contribute to the final image
    (b) Are received by the sodium iodide crystal

(c) Are converted to a voltage signal
(d) Are in the visible spectrum

17. To obtain high-resolution images of a small organ, gland, or joint, which collimator will be most useful?

(a) Low-energy parallel hole
(b) Flat field
(c) Diverging
(d) Pinhole

18. Which instrument should be used to determine the location of a ^{99m}Tc spill?

(a) Geiger–Müller (GM) survey meter
(b) Portable ionization chamber
(c) NaI well counter
(d) Pocket dosimeter

19. A technologist covers the collimator (which is facing the ceiling) with absorbent paper and obtains a uniformity image using a liquid flood source with added ^{99m}Tc. A round cold spot is seen on the image. Subsequently, the technologist removes the collimator, turns the camera face down, and obtains another image using a point source placed on the floor. This image appears uniform. What is the most likely cause of the cold spot on the first image?

(a) Collimator defect
(b) Decoupled photomultiplier tube
(c) Subtle crystal crack
(d) Improper energy peaking

20. If two photons strike a sodium iodide crystal at the same time, what will occur?

(a) Neither event will trigger production of light.
(b) The system will perceive only one event which will contain the summed energy of both photons.
(c) Both events will be correctly perceived.
(d) Only the photon closer to the center of the crystal will be perceived.

21. The time after a scintillation crystal has absorbed a photon before it is able to process another event is called:

    (a) Count rate
    (b) Distortion
    (c) Dead time
    (d) Sensitivity

22. The purpose of the photomultiplier tube is:

    (a) To keep any of the electrical signal which resulted from scatter or background from contributing to the final image
    (b) To convert light into an electrical signal and to magnify that signal
    (c) To convert radioactivity into light
    (d) To filter out photons which strike the crystal from oblique angles

23. The purpose of the pulse height analyzer is:

    (a) To minimize the amount of scattered radiation in the final image
    (b) To convert light into an electrical signal and to magnify that signal
    (c) To convert radioactivity into light
    (d) To filter out photons which strike the crystal from oblique angles

24. A camera with three-pulse height analyzers will be most useful for imaging:

    (a) ^{99m}Tc
    (b) ^{67}Ga
    (c) ^{111}In
    (d) ^{133}Xe

25. A technologist has set a 15% symmetric window for a 140 keV photopeak. What will happen to a 158 keV signal?

    (a) It will be accepted by the pulse height analyzer.
    (b) It will be rejected by the pulse height analyzer.

26. A typical energy spectrum of ^{99m}Tc from a patient contains a broad peak from 90 to 140 keV that is not present in the energy spectrum from a point source of the same isotope in air. This represents:

    (a) The iodine escape peak
    (b) The signal from ^{99m}Tc
    (c) Lead X-ray peak
    (d) Compton scatter

27. Which of the following can be measured in millimeters?

    (a) Energy resolution
    (b) Spatial resolution
    (c) Field uniformity
    (d) Temporal resolution

28. The energy resolution of an image is better with a _____ energy isotope than with a _____ energy isotope.

    (a) Higher, lower
    (b) Lower, higher

29. Of the following types of transmission phantoms, which one requires the fewest images to test the spatial resolution over the entire detector face?

    (a) Hine–Duley phantom
    (b) Four-quadrant bar phantom
    (c) Parallel-line equal-space phantom
    (d) Orthogonal hole phantom

30. According to the curve for ^{137}Cs shown in Fig. 4.1, what is the percent energy resolution of the system?

    (a) 6.9%
    (b) 8.2%
    (c) 10.5%
    (d) 11.8%

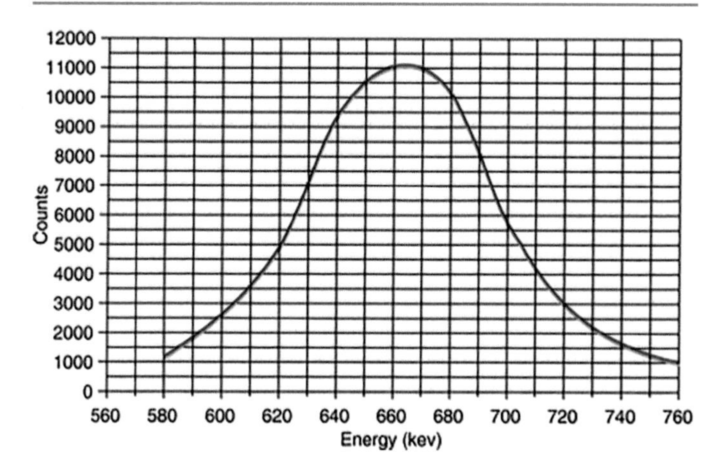

Fig. 4.1 Energy spectrum recorded for ^{137}Cs source in air

31. An orthogonal hole phantom cannot be used to test spatial linearity since it does not contain lead bars.

    (a) True
    (b) False

32. When using a 150 μCi source of ^{99m}Tc to check uniformity, a technologist measures the counts in 1 min as 53,125. If the background in the room is 405 cpm, what is the sensitivity of the system?

    (a) 351 cpm/μCi
    (b) 354 cpm/μCi
    (c) 356 cpm/μCi
    (d) 360 cpm/μCi

33. On Monday, the extrinsic uniformity flood for a gamma camera has a uniformity of 3%. On Tuesday, the uniformity is 8%. The technologist then verifies that the photopeak is still centered in the energy window. What should be the next step?

    (a) Scan a bar phantom instead.
    (b) Call for service.
    (c) Remove the collimator and obtain an intrinsic uniformity flood.

34. Why might a technologist use an asymmetric energy window around a photopeak?

(a) To include other emissions
(b) To exclude edge packing
(c) To exclude Compton scatter
(d) To increase the count rate

35. What energies are included if a 20% symmetric window is used for the 364 keV photopeak of ^{131}I?

(a) 191–437
(b) 328–400
(c) 337–391
(d) 344–384

36. The images acquired as a SPECT camera rotates around a patient are called:

(a) Reconstructions
(b) Arcs
(c) Projections
(d) Azimuth stops

37. With a matrix size of 64 × 64 and two sources placed 30 cm apart, there are 44 pixels between the activity peaks on the activity profile that is generated. What is the pixel size?

(a) 6.8 cm/pixel
(b) 6.8 mm/pixel
(c) 1.4 mm/pixel
(d) 1.4 cm/pixel

38. What steps should be taken to reduce a star artifact from reconstructed SPECT images?

(a) Decrease the matrix size
(b) Increase the time per projection
(c) Increase the radius of projection
(d) Increase the number of projections

39. The _____ method of SPECT acquisition requires slightly more time but has less blurring than the _____ method.

(a) Step and shoot, continuous rotation
(b) Continuous rotation, step and shoot

40. In SPECT imaging, if the image matrix size is increased from 64 × 64 to 128 × 128, which one of the following is true?

    (a) Total imaging time may need to be increased.
    (b) Spatial resolution will decrease.
    (c) Image reconstruction will be faster.
    (d) Image file size will decrease.
    (e) Count rate will decrease.

41. If the distance between two point sources placed on the camera surface is 35 cm in both the x and y directions, and the number of pixels between their activity profile peaks in the image is 52 pixels in the x-axis direction and 55 pixels in the y-axis direction, are the pixel dimensions in the x and y directions within 0.5 mm of one another?

    (a) Yes
    (b) No

42. A technologist measures a standard of known activity in the dose calibrator three times. She subtracts background from each measurement and compares these measurements to the decay-corrected activity of the standard. This was a test of:

    (a) Constancy
    (b) Linearity
    (c) Accuracy
    (d) Geometric dependence

43. Records of dose calibrator checks must be kept for:

    (a) 3 years
    (b) 5 years
    (c) 10 years
    (d) As long as the license is valid

44. In PET imaging, what is the outcome of imaging positrons with a long range of travel before undergoing annihilation?

    (a) Higher rate of random coincidences
    (b) Poorer spatial resolution
    (c) Reduced scatter fraction
    (d) Improved energy resolution

45. Survey meters must be checked for proper operation against a sealed source:

    (a) Before each day of use
    (b) Weekly
    (c) Monthly
    (d) Every 6 months

46. Which of the following would be used to perform a survey of an area in which a small amount of radioactivity is expected to be present?

    (a) Ionization chamber
    (b) Geiger–Müller counter
    (c) Well counter
    (d) Single-probe counting system

47. An ionization chamber with a relatively low applied voltage from the anode to the cathode has a low efficiency and is therefore best for measuring:

    (a) Low exposure rates
    (b) High exposure rates
    (c) Background radiation
    (d) Absorbed dose

48. Most well counters measure radioactivity by the use of:

    (a) An applied voltage from anode to cathode
    (b) A sodium iodide crystal
    (c) A silver halide layer
    (d) Thermoluminescent crystals

49. When measuring samples in a well counter, which one of the following is true?

    (a) Sample volumes should be large.
    (b) Samples should have high radioactivity.
    (c) Sample geometry should be consistent.
    (d) Sample containers do not affect the measurement.

50. A dose calibrator is received in the nuclear medicine department following repair. A linearity test using ^{99m}Tc is performed on the dose calibrator, and the results are as follows:

Measured activity	Date	Time	DF
33.9 mCi	17 Jan	08:00	1
30.3 mCi	17 Jan	09:00	0.891
9.2 mCi	17 Jan	19:00	0.281
2.3 mCi	18 Jan	07:00	0.070
0.50 mCi	18 Jan	20:00	0.016
0.25 mCi	19 Jan	03:00	0.007

What is the next step the technologist should take?

(a) Send the instrument for further repair.
(b) Accept the instrument for use.
(c) Reject the instrument for use.
(d) Change the battery.

51. A check source is calibrated for June 13, 2014, at noon and contains 145 μCi of ^{137}Cs (T1/2 = 30 years). If a dose calibrator is used to measure that dose on June 13, 2020, what is the range of acceptable activities to prove that the accuracy is within 10% (6-year decay factor is 0.871)?

(a) 113–139 μCi
(b) 119–125 μCi
(c) 120–133 μCi
(d) 125–140 μCi

52. 1 mL of a solution is assayed and contains 206 mCi. If the solution is diluted to 9 mL and the activity is now 170 mCi, what is the geometric correction factor?

(a) 0.65
(b) 1.2
(c) 10.9
(d) 18

53. If the energy range for ^{137}Cs at full width half maximum ranges from 623 keV to 701 keV, what is the percent energy resolution?

(a) 7.5%
(b) 10.8%
(c) 11.8%
(d) 13.5%

Image Presentation and Computers

5

1. The type of computer memory that allows temporary storage for programs and data is called:

   (a) ROM
   (b) RAM
   (c) Rad
   (d) REM

2. Assuming that the speed of retrieval is not important, which archiving option would be the best choice for a department with a limited budget?

   (a) Magneto-optical disk
   (b) DVD
   (c) Magnetic tape
   (d) Hard drive

3. A high-pass image filter removes:

   (a) Relatively lower frequencies
   (b) Relatively higher frequencies
   (c) Frequencies that are both too high and too low
   (d) Edges

E. Mantel et al., *Nuclear Medicine Technology*,
https://doi.org/10.1007/978-3-031-26720-8_5

4. Display devices in nuclear medicine include all of the following EXCEPT:
   (a) CRTs
   (b) Video monitors
   (c) Magnetic tape
   (d) Photographic film

5. Which part of the computer is used for data filtering?
   (a) Buffer
   (b) Array processor
   (c) ROM
   (d) ADC

6. Which of the following is NOT true regarding data acquisition in frame mode?
   (a) It requires much less memory than list mode.
   (b) It has a higher acquisition rate compared to list mode.
   (c) Data cannot be divided into shorter images at a later time.
   (d) Data being stored includes periodic clock markers.

7. Which is more useful for gated first-pass cardiac studies?
   (a) List mode
   (b) Frame mode

8. When discussing medical images, what does the acronym "DICOM" stand for?
   (a) Department of Internal Communications
   (b) Digital Imaging and Communications in Medicine
   (c) Digitization Committee
   (d) Division of Computation

9. What is the standard orientation of tomographic DICOM slices?
   (a) Transverse
   (b) Sagittal
   (c) Coronal
   (d) Right anterior oblique

10. Which of the following types of information would NOT be found in a DICOM header?

    (a) Modality, scanner manufacturer, scanner model, FOV
    (b) Pixel size, slice thickness, number of rows and columns
    (c) Patient's insurance carrier, payment method, balance due
    (d) Patient name, ID, birth date, referring physician

11. Which of the following modalities may NOT have separate DICOM files for each slice?

    (a) CT
    (b) MR
    (c) PT
    (d) NM

12. What does the acronym "PACS" stand for?

    (a) Photoconductor array camera and spectrophotometer
    (b) Philanthropic and charitable societies
    (c) Picture archiving and communication system
    (d) Public affairs and community services

13. In information technology, what does the acronym "RIS" stand for?

    (a) Radiology information system
    (b) Research and information system
    (c) Research innovation services
    (d) Regulatory issue statement

14. Filtering of SPECT data may take place:

    (a) Before reconstruction
    (b) During reconstruction
    (c) After reconstruction
    (d) All of the above

15. Which type of filter is used exclusively with dynamic images?

    (a) Spatial
    (b) Temporal
    (c) Band pass
    (d) Low pass

16. Double-emulsion film is used most often in nuclear medicine departments.
    (a) True
    (b) False
17. If a technologist notices an unexpected hot spot on an image, what should he or she NOT do:
    (a) Take oblique or lateral views of the area.
    (b) Attempt to remove the source of the signal by cleansing or removing clothing and reimaging, noting the difference for physician (pants removed, etc.).
    (c) Nothing.
    (d) Survey the area for contamination after the patient leaves.
18. When presenting a bone scan to a physician for interpretation, the technologist should be certain that which of the following information is available?
    (a) Laterality (right vs. left side)
    (b) Injection site
    (c) Age
    (d) Fracture history
    (e) All of the above
19. Film should be stored:
    (a) On its side
    (b) After the foil wrapper has been removed
    (c) In a temperature-controlled area
    (d) All of the above
20. Static on photographic film may result from:
    (a) Removing a sheet of the film too quickly from the box or cassette
    (b) Contact with skin
    (c) Contact with dust and/or lint particles
    (d) All of the above

21. If photographic films appear too light, which of the following should be monitored?

    (a) Chemicals are replenished at the correct rate.
    (b) Developer temperature is set according to the manufacturer's recommendation.
    (c) Water does not overflow into the developer in the film processor.
    (d) All of the above.

22. Which of the following describes the fixing process?

    (a) The undeveloped sodium iodide crystals are removed, and the developing process is stopped.
    (b) Silver halide crystals become reduced to metallic silver.
    (c) The undeveloped silver halide crystals are removed, and the developer is neutralized.
    (d) None of the above.

23. Filtered back projection of SPECT data initially creates:

    (a) Transverse images
    (b) Sagittal images
    (c) Coronal images
    (d) Oblique images

24. Which of the following is NOT a characteristic used to evaluate image quality in NM images?

    (a) Gray scale
    (b) Spatial resolution
    (c) Contrast
    (d) Noise

25. When placing a region of interest (ROI) on an image in order to extract numerical data, which of the following is false?

    (a) ROIs must be accurately placed on the tissue of interest.
    (b) ROIs must be circular.
    (c) ROIs can be defined automatically using edge detection algorithms.
    (d) When ROIs are copied across a time series of images, a time-activity curve can be generated.

26. Which of the following are NOT statistics that can be obtained from an ROI?

    (a) Minimum and maximum pixel values
    (b) Mean pixel value
    (c) Standard deviation of the pixel values
    (d) Contrast resolution
    (e) Area

Skeletal System Scintigraphy

6

1. Which of the following is a malignant bone disease?
   (a) Paget's disease
   (b) Ewing's sarcoma
   (c) Osteomyelitis
   (d) Osteoid osteoma
2. A three-phase bone scan is often done to differentiate
   (a) Osteoporosis vs. cellulitis
   (b) Osteomyelitis vs. diskitis
   (c) Osteomyelitis vs. cellulitis
   (d) Osteoporosis vs. septic arthritis
3. The presence of gastric and thyroid activity on a bone scan signals the presence of:
   (a) Metastatic disease
   (b) Free pertechnetate
   (c) Radionuclide impurity
   (d) Reducing agent
4. What is the purpose of a reducing agent in a ^{99m}Tc diphosphonate kit?
   (a) To oxidize technetium
   (b) To lower the valence state of technetium
   (c) To improve the tag efficiency

E. Mantel et al., *Nuclear Medicine Technology*,
https://doi.org/10.1007/978-3-031-26720-8_6

(d) Both (b) and (c)
(e) (a), (b), and (c)

5. What is the dose of ^{99m}Tc MDP most often prescribed for a planar bone scan?

(a) 1–3 mCi
(b) 5–10 mCi
(c) 20–30 mCi
(d) 30–35 mCi

6. Which of the following is least likely to cause an artifact on bone scan?

(a) Snap on trousers
(b) Colostomy bag
(c) Skin contaminated by urine
(d) Injection site

7. What is not an indication for a bone scan?

(a) Metastatic disease
(b) Osteoporosis
(c) Cellulitis
(d) Avascular necrosis

8. The presence of free pertechnetate on a bone scan may be the result of:

(a) The use of a radiopharmaceutical which was prepared too long ago
(b) Introduction of air into the kit vial while adding technetium
(c) Increased blood flow to bones
(d) Both (a) and (b)
(e) (a), (b), and (c)

9. What is the purpose of hydration and voiding after an injection for a bone scan?

(a) To block the uptake of unlabeled technetium by the stomach
(b) To reduce the possibility of urine contamination
(c) To obtain a superscan
(d) To reduce the radiation dose to the bladder

10. What could be the cause of generalized, diffuse activity in the abdomen on a bone scan?

    (a) Free pertechnetate
    (b) Malignant ascites
    (c) Pacemaker
    (d) Bone cyst

11. What timing protocol best describes a four-phase bone scan?

    (a) During injection, immediately following injection, 2–4 h, and 18–24 h
    (b) During injection, 2–4 h, 24 h, and 48 h
    (c) During injection, immediately following injection, 2–4 h, and 6 h
    (d) None of the above

12. A focal hot spot near the left femur shows up on a bone scan. What is/are the best way/ways to proceed?

    (a) Change to pinhole collimator and image.
    (b) Perform SPECT imaging.
    (c) Have the patient remove clothing of that area.
    (d) Ask the patient to wash the skin in that area with soap and water.
    (e) (c) and (d).

13. A bone scan showing relatively uniformly increased skeletal uptake of radiopharmaceutical with almost absent renal and bladder activity is usually referred to as a:

    (a) Flare phenomenon
    (b) Superscan
    (c) Renal failure
    (d) Suprascan

14. The glove phenomenon is usually the result of:

    (a) Reactive arthritis
    (b) Intravenous injection
    (c) Antecubital injection
    (d) Arterial injection
    (e) Subcutaneous injection

15. What are common sites of bony metastasis?

    (a) Pelvis
    (b) Spine
    (c) Ribs
    (d) All of the above

16. What is the purpose of a stannous ion in a diphosphonate kit?

    (a) Acts as a reducing agent
    (b) Acts as an oxidizing agent
    (c) Provides a stabilizing force
    (d) Maintains particle size

17. The appendicular skeleton includes the following bones, except:

    (a) The femurs
    (b) The skull
    (c) The phalanges
    (d) The radius

18. The axial skeleton contains:

    (a) The ribs
    (b) The skull
    (c) The vertebral column
    (d) The pelvis
    (e) All of the above
    (f) (a), (b), and (c) only

19. By what mechanism do diphosphonates localize in the bone?

    (a) Capillary blockade
    (b) Active transport
    (c) Ion exchange
    (d) Phagocytosis

20. Which of the following describes a pediatric bone scan?

    (a) Increased uptake in long bones
    (b) Decreased uptake along epiphyseal plates
    (c) Increased uptake along epiphyseal plates
    (d) Overall decreased uptake in the bone

21. The first phase of a three-phase bone scan is best performed by:

    (a) Bolus injection followed by dynamic 2-s images for 60 s
    (b) Bolus injection followed by dynamic 20-s images for 3 min
    (c) Bolus injection followed by a static 500–600-K count image
    (d) Bolus injection followed by dynamic 1-s images for 30 s

22. What pharmaceuticals may be used for bone marrow imaging:

    (a) ^{99m}Tc albumin colloid
    (b) ^{99m}Tc sulfur colloid
    (c) ^{99m}Tc PYP
    (d) (a) and (b)
    (e) All of the above

23. What is often used in imaging-suspected avascular necrosis of the hip?

    (a) SPECT imaging
    (b) PET imaging
    (c) Pinhole collimation
    (d) Diverging collimation
    (e) Both (a) and (c)

24. Splenic uptake on a bone scan is often associated with

    (a) Liver failure
    (b) Sickle cell disease
    (c) Splenic abscess
    (d) Paget's disease

25. The bone is made of:

    (a) Hydrogen peroxide
    (b) Hydroxyapatite mineral
    (c) Collagen
    (d) (a) and (b) only
    (e) (b) and (c) only

26. Osteoblastic activity refers to:

    (a) Bone marrow biopsy
    (b) Destruction and reabsorption of the bone
    (c) Bone compression
    (d) New bone formation

27. Osteoclastic activity refers to:

    (a) Bone marrow biopsy
    (b) Destruction and reabsorption of the bone
    (c) Bone compression
    (d) New bone formation

28. The function of the skeleton is:

    (a) To provide support
    (b) To protect organs
    (c) To produce blood cells
    (d) All of the above
    (e) (a) and (b) only

29. Which group shows the highest rate of primary bone tumors?

    (a) The elderly
    (b) Children
    (c) Males
    (d) Females

30. The radiation dose from a bone scan is highest to the:

    (a) Bone marrow
    (b) Chest
    (c) Bladder
    (d) Brain

31. When performing a bolus injection for a three-phase bone scan, why would the tourniquet be released and injection delayed for 1 min?

    (a) To minimize pain during injection
    (b) To reduce transient hyperemia resulting from vasodilation
    (c) To double-check the dynamic sequence settings
    (d) To obtain a better bolus

32. What is an advantage of spot planar imaging over whole-body imaging for a bone scan?

    (a) Speed.
    (b) Decreased patient-to-detector distance.
    (c) Less tracer is used.
    (d) No need for COR correction.

33. Which of the following would be a reason not to inject in the right antecubital fossa?

    (a) The patient has had blood drawn from the back of the right hand that same day.
    (b) The patient was injected on the right for a previous bone scan.
    (c) There is a suspicion of osseous abnormality in the right distal humerus.
    (d) The patient is left-handed.

34. The preparation for a bone scan is:

    (a) NPO from midnight
    (b) Cleansing enema
    (c) The patient to be off thyroid medication for 4 weeks
    (d) None of the above
    (e) (a) and (b) only
    (f) All of the above

35. Metastases usually affect the axial skeleton before the appendicular skeleton.

    (a) True
    (b) False

36. The advantage(s) of bone scanning over plain radiography is (are):

    (a) The bone must lose only a minimum of calcium content before lesions are visible on bone scintigraphy.
    (b) Time.
    (c) Efficient for multifocal trauma such as child abuse.
    (d) All of the above.
    (e) (a) and (c) only.

37. The mechanism of localization for bone marrow scanning is:

    (a) Active transport
    (b) Ion exchange
    (c) Phagocytosis
    (d) Capillary blockade

38. A large amount of diffuse soft tissue activity present on a bone scan at 4 h frequently represents:

    (a) Increased cardiac output
    (b) Renal insufficiency
    (c) Metastatic disease
    (d) Infection

39. What imaging agent can be used to image the skeleton as well as the heart?

    (a) ^{99m}Tc MDP
    (b) ^{99m}Tc HDP
    (c) ^{99m}Tc PYP
    (d) ^{201}TL chloride

40. Rib fractures often show up as:

    (a) Multiple focal hot spots located in consecutive ribs
    (b) A linear increased activity along the long axis of the rib
    (c) Diffuse activity in the chest cavity
    (d) Cold spots

41. The cold defect in the left proximal humerus in Fig. 6.1 is most likely the result of:

    (a) Shielding
    (b) Pacemaker
    (c) A bandage around the left upper arm
    (d) Surgically implanted metal
    (e) Motion

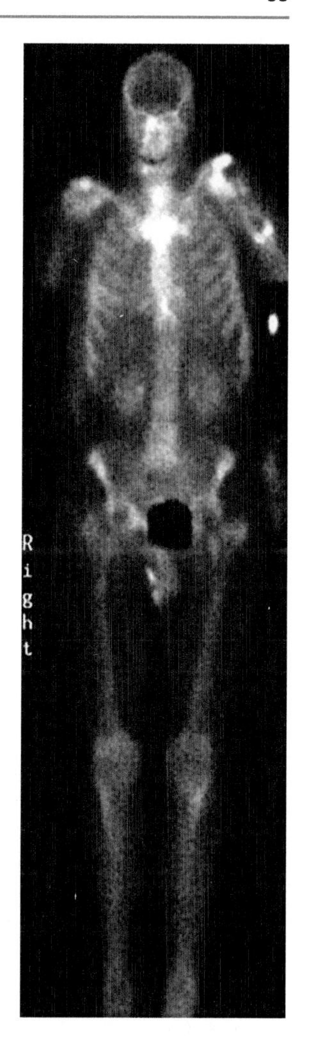

Fig 6.1 Whole body bone scan anterior view

Central Nervous System Scintigraphy

7

1. The venous phase of a cerebral blood flow study is signaled by:
   (a) Visualization of the jugular veins
   (b) Appearance of the radiopharmaceutical in the superior sagittal sinus
   (c) Disappearance of radiopharmaceutical from the carotid arteries
   (d) Appearance of radiopharmaceutical in the middle cerebral arteries
2. The localization of ^{99m}Tc HMPAO is related to:
   (a) A breakdown of the blood–brain barrier
   (b) Cerebral blood flow
   (c) Glucose metabolism
   (d) Distribution of neuroreceptors
3. CSF is made mostly of:
   (a) Protein
   (b) Water
   (c) Glucose
   (d) Blood
4. The function of cerebrospinal fluid is to:
   (a) Protect the brain and spinal cord against shock
   (b) Filter plasma

E. Mantel et al., *Nuclear Medicine Technology*,
https://doi.org/10.1007/978-3-031-26720-8_7

 (c) Produce neurotransmitters
 (d) All of the above

5. When positioning for a radionuclide angiogram, the patient should be positioned:
 (a) Posteriorly
 (b) Anteriorly
 (c) With as much facial activity as possible excluded
 (d) None of the above

6. Exametazime is also known as:
 (a) DTPA
 (b) MAG3
 (c) HMPAO
 (d) ECD

7. Bicisate is also known as:
 (a) DTPA
 (b) MAG3
 (c) HMPAO
 (d) ECD

8. CSF dynamics are studied following administration of ^{111}In DTPA:
 (a) Via intravenous injection
 (b) Via intraperitoneal injection
 (c) Via intrathecal injection
 (d) Via intradermal injection

9. The dose which would commonly be used for a radionuclide angiogram with ^{99m}Tc DTPA is:
 (a) 5 mCi
 (b) 10 mCi
 (c) 20 mCi
 (d) 25 mCi

10. Which of the following is true regarding injection of the radiopharmaceuticals for SPECT brain imaging?

    (a) Lights should be dimmed during injection.
    (b) The patient should be encouraged to read or watch TV during injection to take his or her mind off any pain.
    (c) Family members should be allowed to talk to the patient during injection.
    (d) Injection should be made immediately after venous puncture to avoid clot formation.

11. Which dose could be used for ^{99m}Tc ECD or ^{99m}Tc HMPAO for SPECT brain imaging?

    (a) 2–5 mCi
    (b) 5–7 mCi
    (c) 7–10 mCi
    (d) 20 mCi

12. Ictal SPECT and PET studies require an injection of radiopharmaceutical:

    (a) During a seizure
    (b) Immediately after a seizure
    (c) Between seizures
    (d) Both (a) and (b)

13. HMPAO and ECD are lipophilic agents that do not significantly redistribute in the brain.

    (a) True
    (b) False

14. Advantages of using ECD over HMPAO for SPECT imaging of the brain include which of the following?

    (a) Can be injected up to 6 h after preparation
    (b) More rapid clearance from the bloodstream
    (c) Better target-to-background ratio
    (d) All of the above
    (e) None of the above

15. White matter has about four times greater blood flow than gray matter.

    (a) True
    (b) False

16. Planar images of the brain using ^{99m}Tc DTPA are obtained:

    (a) Immediately after the dynamic exam
    (b) 30 min after injection
    (c) 1 h after injection
    (d) 1.5 h after injection

17. The purpose of placing an elastic band around a patient's head just above the orbits during a radionuclide angiogram is:

    (a) To decrease the activity from the orbits
    (b) To increase intracranial pressure
    (c) To decrease blood supply to small scalp vessels
    (d) To immobilize the patient

18. The images obtained with a PET study using ^{18}F FDG can best be described as:

    (a) Anatomical
    (b) Perfusion
    (c) Metabolic
    (d) Dynamic

19. In a Diamox challenge study, vascular disease will appear as decreased perfusion after the use of Diamox.

    (a) True
    (b) False

20. The injection for a CSF shunt patency study is:

    (a) Intravenous
    (b) Subcutaneous
    (c) Intrathecal
    (d) Into the shunt reservoir or tubing

21. Shunts which divert the flow of CSF are often used to treat:

    (a) Communicating hydrocephalus
    (b) Noncommunicating hydrocephalus
    (c) CSF leak
    (d) All of the above
    (e) (a) and (b) only

22. Which of the following agents will cross an intact blood–brain barrier?

    (a) ^{99m}Tc DTPA
    (b) ^{99m}Tc GH
    (c) ^{99m}Tc pertechnetate
    (d) ^{99m}Tc HMPAO

23. Which radiopharmaceutical is not commonly used for brain perfusion imaging with SPECT?

    (a) ^{99m}Tc DTPA
    (b) ^{99m}Tc HMPAO
    (c) ^{99m}Tc ECD

24. Placing an elastic band around a patient's head just above the orbits during a radionuclide brain death scan is recommended for which of the following?

    (a) ^{99m}Tc DTPA
    (b) ^{99m}Tc HMPAO
    (c) ^{99m}Tc ECD

25. Radionuclide angiography is most commonly used to:

    (a) Locate neoplasms
    (b) Investigate brain perfusion
    (c) Evaluate cerebrovascular reserve
    (d) Confirm brain death

26. In which area is the radiopharmaceutical not visible in a normal cisternogram?

    (a) Basal cisterns
    (b) Cerebral convexities

(c) Lateral ventricles
(d) Sylvian fissure

27. What is the usual dose for a radionuclide cisternogram in adults?

(a) 5 mCi of ^{111}In DTPA
(b) 100 μCi of ^{111}In DTPA
(c) 500 μCi of ^{111}In DTPA
(d) 2.5 mCi of ^{111}In DTPA

28. What is the purpose of pledget placement in a suspected CSF leak?

(a) To make the patient more comfortable during subarachnoid puncture
(b) To alter the biodistribution of the radiotracer
(c) To test them for contamination suggesting the presence of leaking CSF in the nose or ears
(d) To avoid contamination of the scintillation detection system

29. Why is ^{111}In DTPA preferred over ^{99m}Tc DTPA for cisternography in adults?

(a) Because it has lower energy photons
(b) Because it has two photopeaks
(c) Because it has a longer half-life allowing delayed imaging
(d) Because it has a superior biodistribution

30. Considerations for successful SPECT imaging of the brain include:

(a) Minimizing sensory stimulation during injection
(b) Minimizing patient-to-detector distance
(c) Immobilizing the patient
(d) All of the above
(e) (b) and (c) only

31. To prepare a patient for DaTscan® study, which of the following is correct?
    (a) NPO overnight or at least 4 h
    (b) No coffee or tea at least 12 h prior to the study
    (c) Thyroid blockage by SSKI prior to the study is required
    (d) Void before imaging

32. For DaTscan® study, which of the following is correct?
    (a) DaTscan® detects beta-amyloid deposition.
    (b) DaTscan® detects brain infarction.
    (c) DaTscan® detects brain presynaptic dopamine transporter.
    (d) DaTscan® detects brain neurofibrillary tangles.
    (e) DaTscan® detects brain Lewy bodies.

33. To perform a DaTscan® study, which of the following is correct?
    (a) Imaging can be obtained 3–6 h after tracer injection.
    (b) Dedicated head holder should be used.
    (c) The camera should be very close to the head.
    (d) (a–c).

34. For Amyvid® PET study, which of the following is correct?
    (a) NPO for 4 h before the study is needed.
    (b) Blood sugar should be <200 mg/dl.
    (c) Arms should be above shoulders during imaging.
    (d) Imaging can be started as soon as 30 min after tracer injection.
    (e) None of the above.

35. For Amyvid® PET study, which of the following is correct?
    (a) Amyvid® detects brain ischemia.
    (b) Amyvid® detects brain atrophy.
    (c) Amyvid® detects brain beta-amyloid deposition.
    (d) Amyvid® detects brain neurofibrillary tangles.
    (e) Amyvid® detects brain Lewy bodies.

36. To prepare a patient for FDG-PET of the brain, which of the following is correct?
    (a) NPO at least 4 h.
    (b) No coffee or tea at least 12 h prior to the study.
    (c) Blood sugar should be <200 mg/dl.
    (d) Imaging can be started as soon as 30 min after tracer injection.
    (e) All of the above.

37. Amyloid PET imaging is most helpful in patients with which of the following conditions?
    (a) Suspected Parkinson's disease but without definite diagnosis
    (b) Suspected Alzheimer's disease but without definite diagnosis
    (c) Clinically diagnosed Parkinson's disease, to confirm with imaging findings
    (d) Clinically diagnosed Alzheimer's disease, to confirm with imaging findings

38. Radiotracers that have been approved by the FDA for imaging Aβ plaques include all of the following except:
    (a) 18F-florbetapir (Amyvid®)
    (b) 18F-flutemetamol (Vizamyl®)
    (c) 18F-florbetaben (NeuraCeq™)
    (d) 11C-Pittsburgh Compound B (PIB)

39. For Amyloid PET study, a patient should:
    (a) Be on NPO overnight
    (b) Hold medicines on the day of imaging
    (c) Hold medicines working on the central nervous system only, on the day of imaging
    (d) Have no special preparations regarding diet and medicine

40. Which of the following is not correct regarding the recommended dose and waiting period before amyloid PET image acquisition?

    (a) 18F-florbetapir, dose of 370 MBq (10 mCi), waiting period of 45 min
    (b) 18F-flutemetamol, dose of 185 MBq (5 mCi), waiting period of 90 min
    (c) 18F-florbetapir, dose of 370 MBq (10 mCi), waiting period of 120 min
    (d) 18F-florbetaben, dose of 300 MBq (8 mCi), waiting period of 120 min

41. Amyloid PET is indicated in patients with the following conditions *except*:

    (a) Persistent or progressive MCI with uncertain clinical diagnosis
    (b) Diagnosed Alzheimer's disease, to determine the severity of dementia
    (c) Clinically suspected possible Alzheimer's disease but with atypical clinical course
    (d) Clinically progressive dementia with an early onset (65 years or less in age)

42. Which of the following tracer is used for evaluation of memory disorder:

    (a) 18F-FDG
    (b) 18F-fluciclovine
    (c) 18F-NaF
    (d) 68Ga-DOTATATE

Cardiovascular System Scintigraphy

8

1. What portion of an ECG wave represents the depolarization of the ventricles?

   (a) QRS complex
   (b) P wave
   (c) ST segment
   (d) T wave

2. Contraindication(s) to treadmill exercise cardiac stress testing include:

   (a) Acute MI (within 3–4 days)
   (b) Uncontrolled hypertension (SBP >210 mmHg and/or DBP >110 mmHg)
   (c) Acute illness for any reason
   (d) Patient on beta-blocker

3. Within what time frame must imaging be initiated following injection of ^{201}Tl chloride in an exercise perfusion study?

   (a) Within 5 min
   (b) Within 10–15 min
   (c) Within 20–30 min
   (d) Within 4 h

E. Mantel et al., *Nuclear Medicine Technology*, https://doi.org/10.1007/978-3-031-26720-8_8

4. For SPECT myocardial perfusion imaging, which of the following is not correct:

   (a) With ^{99m}Tc sestamibi or ^{99m}Tc tetrofosmin, imaging should be begun approximately 45 min after stress testing.
   (b) With ^{201}Tl, imaging should be begun approximately 10 min after stress testing.
   (c) With ^{99m}Tc sestamibi or ^{99m}Tc tetrofosmin, imaging acquisition after stress testing can be repeated multiple times when patient motion or other artifacts are suspected.
   (d) With ^{201}Tl, imaging acquisition after stress testing can be repeated multiple times when patient motion or other artifacts are suspected.

5. Of the choices offered, which is the best imaging view for calculating left ejection fraction?

   (a) Anterior
   (b) LAO 5°
   (c) LAO 45°
   (d) Left lateral

6. Given the data below, calculate the left ventricular ejection fraction.

   Net ED = 58,219
   Net ES = 35,317

   (a) 16%
   (b) 39%
   (c) 43%
   (d) 60%

7. The left ventricular ejection fraction determined in Question 6 is within the normal range.

   (a) True
   (b) False

8. Figure 8.1 shows computer-generated time-activity curves over a cardiac cycle for two patients. Which curve represents the patient with the higher ejection fraction?

   (a) 1a
   (b) 1b

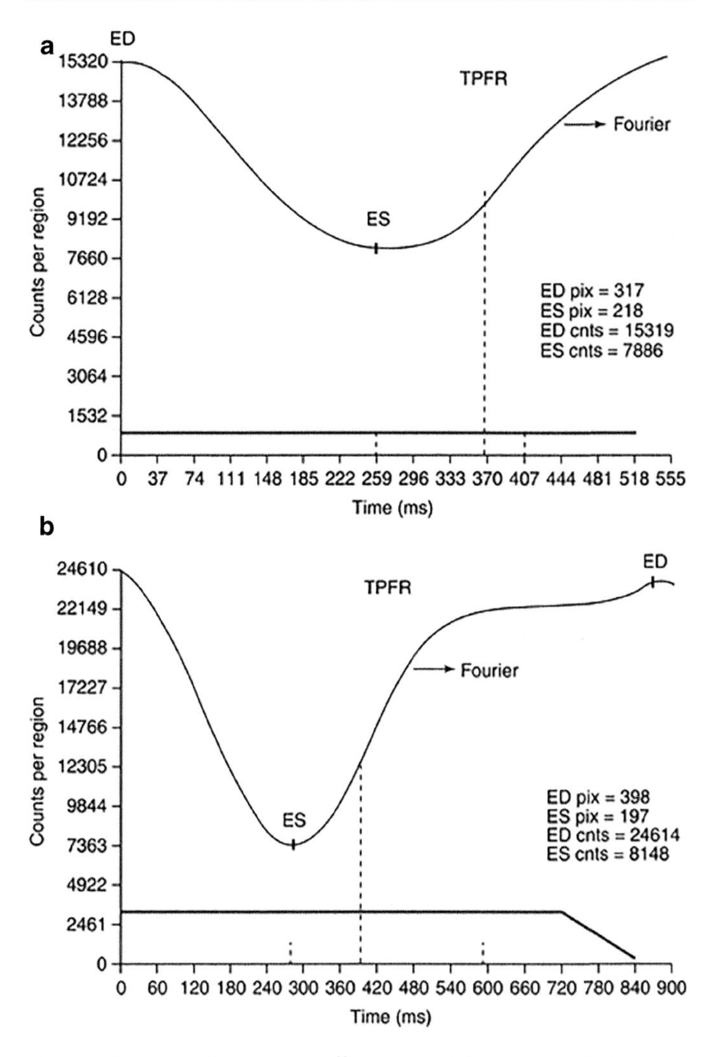

Fig. 8.1 Time-activity curves for ^{99m}Tc sestamibi cardiac perfusion gated imaging

9. Which of the following cannot be evaluated by radionuclide ventriculography?
   (a) Wall motion
   (b) Wall thickness
   (c) Aneurysmal flow
   (d) Conduction abnormalities

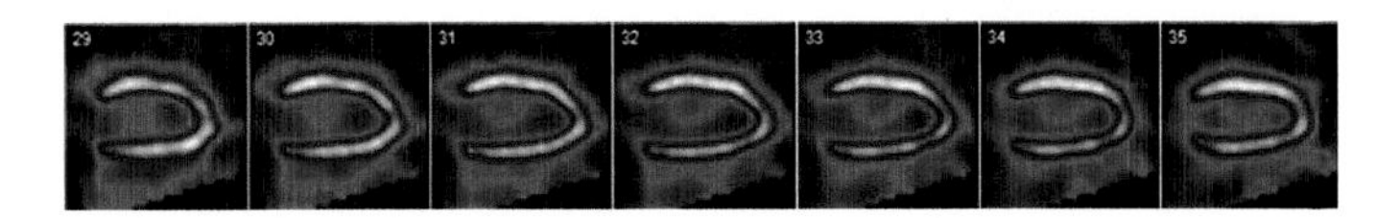

Fig. 8.2 SPECT imaging of ^{99m}Tc sestamibi cardiac perfusion

10. The images in Fig. 8.2 represent what type of images?
    (a) Short axis
    (b) Vertical long axis
    (c) Horizontal long axis
    (d) None of the above

11. During a gated study, 24 images per cardiac cycle are obtained. If the patient's heart rate is 65 bpm, the length of time per image is:
    (a) 3.8 ms
    (b) 38 ms
    (c) 4.1 ms
    (d) 41 ms

12. For sestamibi SPECT myocardial perfusion imaging, a delay in imaging acquisition (approximately 45 min or a little more after stress testing and tracer injection) has the following benefit except for:
    (a) To allow more tracer accumulation within the myocardium
    (b) To allow the patient to recover from exercise with a more peaceful respiration
    (c) To allow heart rate to return to baseline and more regular cardiac rhythm
    (d) To minimize interference from hepatic uptake

13. What will reverse the effects of dipyridamole?
    (a) Dobutamine
    (b) Cholecystokinin

(c) Aminophylline
(d) Adenosine
(e) Thallium

14. Which of the following will most negatively affect radionuclide ventriculography?

(a) Sinus tachycardia
(b) Sinus bradycardia
(c) Long Q–T interval
(d) Occasional PVC
(e) Sinus arrhythmia

15. The vertical long-axis view of the heart is most comparable to a:

(a) Coronal image
(b) Sagittal image
(c) Transverse image

16. As an alternative to exercise studies, pharmacologic stress can be achieved through the use of any of the following except:

(a) Dipyridamole
(b) Aminophylline
(c) Dobutamine
(d) Adenosine

17. In multigated analysis of the left ventricle, end systole is best described as:

(a) The frame with the highest number of counts in the ROI around the left ventricle
(b) The frame with the lowest number of counts in the ROI around the right ventricle
(c) The frame with the lowest number of counts in the ROI around the left ventricle, plus the counts from the background ROI
(d) All of the above
(e) None of the above

18. All of the following will negatively affect a non-gated myocardial study with ^{201}Tl chloride except:

    (a) Wrong collimator
    (b) Arrhythmia
    (c) Incorrect COR
    (d) Increased patient-to-detector distance
    (e) None of the above

19. Dipyridamole is supplied to a nuclear medicine department in 10 ml vials, each containing 50 mg. If a patient weighs 155 lb., how many milliliters must be injected for him or her to receive 0.56 mg/kg?

    (a) 7.9 mL
    (b) 8.7 mL
    (c) 17.4 mL
    (d) 39.5 mL

20. Which of the following does not describe correct preparation for treadmill exercise myocardial perfusion study?

    (a) NPO at least 4 h prior to radiopharmaceutical administration
    (b) Halt vasodilator medication 24 h before study
    (c) Halt beta-blocker 24 h before study
    (d) No coffee or tea 24 h before study
    (e) b and d

21. During an exercise-gated study, the ECG pattern suddenly becomes dramatically different, although the patient is responsive, has an unchanged pulse rate, continues to exercise, and has no pain. The technologist should first:

    (a) Start CPR
    (b) Call a code
    (c) Check for a disconnected lead
    (d) Call the referring physician
    (e) Do nothing

22. Which of the following radiopharmaceuticals is not used to study the heart?

    (a) ^{99m}Tc sestamibi
    (b) ^{123}I MIBG

(c) ^{99m}Tc PYP
(d) ^{32}P chromic phosphate
(e) Both (b) and (d)

23. Indications for early termination of exercise include:

(a) Moderate-to-severe angina pectoris
(b) Marked dyspnea or fatigue
(c) Dizziness or near-syncope
(d) Patient's request to terminate the test
(e) All of the above

24. For pharmacologic stress tests, which of the following has the shortest half-life?

(a) Adenosine
(b) Dipyridamole
(c) Regadenoson
(d) Dobutamine

25. Performing a MUGA scan to evaluate LV ejection fraction, including the spleen, within the ROI for the background will cause:

(a) Artificially high EF
(b) Artificially low EF
(c) Any changes depend on the activity of the LV
(d) Either b or c

26. ^{82}Rb chloride is the daughter of which isotope?

(a) ^{82}Sr
(b) ^{85}Sr
(c) ^{85}Rb
(d) None of the above

27. Cardiac contraction is initiated in the:

(a) SA node
(b) AV node
(c) Bundle of His
(d) P wave

28. Treadmill exercise tests increase heart rate by:

    (a) Increasing the slope of the treadmill
    (b) Increasing treadmill speed
    (c) Simulating infarct
    (d) All of the above
    (e) (a) and (b) only

29. A high-resolution collimator would be most appropriate for:

    (a) Myocardial perfusion study with ^{201}Tl chloride
    (b) Myocardial perfusion with ^{99m}Tc sestamibi
    (c) First-pass study with ^{99m}Tc DTPA
    (d) Myocardial imaging with ^{99m}Tc pyrophosphate

30. What is the half-life of ^{82}Rb?

    (a) 110 min
    (b) 75 s
    (c) 75 min
    (d) 2 min

31. How is ^{82}Rb produced?

    (a) Nuclear reactor
    (b) Generator
    (c) Cyclotron

32. If a SPECT myocardial study with ^{99m}Tc sestamibi is performed using 64 views of 20 s each over 360°, approximately how long will the study take assuming continuous rotation?

    (a) 10 min
    (b) 20 min
    (c) 30 min
    (d) 40 min

33. Gated equilibrium cardiac blood pool images can be used to reliably determine:

    (a) Hibernating myocardium
    (b) LVEF
    (c) Interventricular shunt

(d) Ischemia
(e) All of the above

34. Akinesis refers to:

(a) A lack of wall motion
(b) Diminished wall motion
(c) Paradoxical wall motion
(d) Septal motion

35. The R–R interval represents:

(a) Only repolarization
(b) Arrhythmia filtering
(c) The length of cardiac cycles
(d) Length of time data held in a buffer before being accepted or rejected

36. When labeling RBCs with ^{99m}Tc for radionuclide ventriculography, the highest labeling efficiency will be obtained by:

(a) The in vivo method
(b) The in vitro method
(c) The modified in vivo/in vitro method
(d) All resulting in the same labeling efficiency

37. What is the usual dose of Rb chloride?

(a) 15 mCi for each of the stress and rest doses
(b) 30 mCi total for the stress and rest scans
(c) 30–60 mCi for each of the stress and rest doses
(d) 15 mCi total for the stress and rest scans

38. How does an infarct appear on a PET perfusion scan?

(a) Focal areas of increased uptake
(b) Diffuse areas of increased uptake
(c) Focal areas of decreased/low uptake
(d) No visual difference

39. Gated blood pool ventriculography is often used to:

(a) Diagnose arrhythmia
(b) Obtain a baseline measurement of cardiac function in potential radiation therapy patients

(c) Detect hematologic spread of metastases
(d) Determine the effect of chemotherapy on cardiac function

40. Setup for a SPECT imaging of myocardial perfusion with ^{99m}Tc sestamibi includes all of the following except:

(a) Ensuring that the detector head will not snag IV lines or oxygen tubing
(b) Setting a 10° caudal tilt to differentiate the left atrium from left ventricle
(c) Moving patient's arms over his or her head
(d) Ensuring that COR correction for high-sensitivity collimator is selected

41. A patient who cannot exercise and who has asthma can undergo a stress myocardial perfusion study through the use of:

(a) The Bruce protocol
(b) Dipyridamole
(c) Dobutamine
(d) Adenosine

42. If performed on the same day, which of the following is not correct?

(a) To perform FDG-PET/CT and then ^{82}Rb PET/CT on the same day
(b) To perform ^{82}Rb PET/CT and then FDG-PET/CT on the same day
(c) To perform sestamibi SPECT and then FDG-PET/CT on the same day
(d) (a) and (b)

43. Which radiopharmaceutical can be used to assess myocardial perfusion, right ventricle ejection fraction, and left ventricle ejection fraction with a single injection?

(a) ^{201}Tl chloride
(b) ^{99m}Tc PYP
(c) ^{99m}Tc-labeled RBCs
(d) ^{99m}Tc sestamibi

44. Choose the correct order in which the structures appear during a first-pass study:

    (a) Right ventricle, pulmonary artery, lungs, pulmonary veins, left ventricle
    (b) Right ventricle, pulmonary veins, lungs, pulmonary artery, left atrium
    (c) Left ventricle, pulmonary artery, lungs, pulmonary veins, right ventricle
    (d) Left ventricle, pulmonary veins, lungs, pulmonary artery, right ventricle

45. Which of the following is the most likely cause of false anterior or septal wall defects on myocardial perfusion imaging with ^{99m}Tc sestamibi or ^{201}Tl chloride?

    (a) Center of rotation error
    (b) Respiratory motion
    (c) Too few projections
    (d) Attenuation by breast tissue

46. During a 1-day myocardial perfusion study with ^{99m}Tc sestamibi including both stress and rest:

    (a) Rest images must be obtained before stress images.
    (b) The second dose of sestamibi should be higher than the first.
    (c) The higher dose of sestamibi is always injected for the stress portion.
    (d) The sestamibi doses should be identical for the best comparison of stress and rest.

47. Why is it necessary to inject two doses of ^{99m}Tc sestamibi for a myocardial perfusion study including both stress and rest conditions?

    (a) Because the physical half-life of ^{99m}Tc is too short to obtain both image sets from one injection.
    (b) Because there is no redistribution of ^{99m}Tc sestamibi within the myocardium.
    (c) Because the effective half-life is too short.
    (d) Because sestamibi is not retained by the myocardium for a sufficient time.

48. Soft tissue attenuation is most problematic in:
    (a) SPECT myocardial imaging with ^{99m}Tc sestamibi
    (b) SPECT myocardial imaging with ^{201}Tl chloride
    (c) SPECT/CT myocardial imaging with ^{99m}Tc sestamibi
    (d) ^{82}Rb PET/CT

49. The bull's-eye display created after myocardial perfusion imaging represents:
    (a) Only short-axis images
    (b) An entire set of stress or rest SPECT images
    (c) All part of the myocardial wall except the apex
    (d) None of the above
    (e) (b) and (c)

50. Differences between myocardial perfusion imaging with ^{201}Tl chloride and ^{99m}Tc sestamibi include:
    (a) More hepatic and GI activity with ^{201}Tl
    (b) Less soft tissue attenuation with ^{201}Tl
    (c) Higher photon flux with ^{99m}Tc
    (d) Less respiratory motion with ^{201}Tl due to delay after exercise before imaging

51. When performing a radionuclide study to evaluate lower extremity lymphedema, what should be injected, and where?
    (a) ^{99m}Tc MAA, in bilateral pedal veins
    (b) ^{99m}Tc MAA, subcutaneously in each foot
    (c) ^{99m}Tc sulfur colloid, in bilateral pedal veins
    (d) ^{99m}Tc sulfur colloid, intradermally, in each foot

52. Following injection of ^{99m}Tc sestamibi, during a resting myocardial perfusion study, imaging is delayed for 1 h because:
    (a) Respiratory motion will decrease.
    (b) Liver and lung activity is too high to allow high-quality images of the myocardium.
    (c) Until 1 h after injection, sestamibi has not had sufficient time to be taken up by the myocardium.
    (d) The patient should be monitored for adverse reactions during the first hour.

53. When performing a dual-isotope myocardial perfusion rest/ stress study, which isotope should be injected first?

    (a) ^{201}Tl.
    (b) ^{99m}Tc.
    (c) It does not matter which is injected first, as long as ^{99m}Tc is used for the stress study.
    (d) Either can be injected first without any consequence.

54. The advantage of ^{82}Rb PET cardiac perfusion study (vs. SPECT imaging) includes all of the following except:

    (a) Better imaging resolution
    (b) Less radiation exposure
    (c) Ideal for overweight patients
    (d) Able to determine myocardial ischemia and viability

55. False-negative stress myocardial perfusion studies are usually caused by:

    (a) Failure of the myocardium to extract ^{99m}Tc sestamibi from the bloodstream
    (b) Electrocardiogram leads
    (c) Failure of the patient to reach maximal stress
    (d) ^{99m}Tc sestamibi contained in a too large volume

56. Which of the following should be prepared ahead of a dipyridamole stress myocardial perfusion study?

    (a) Dose of dipyridamole
    (b) Dose of radiopharmaceutical
    (c) Dose of aminophylline
    (d) All of the above
    (e) (a) and (b) only

57. During a MUGA study, data collection is stopped when:

    (a) A preset number of counts have been acquired.
    (b) A preset number of cardiac cycles have been reached.
    (c) A preset time has been reached.
    (d) Either a preset number of counts or a preset number of cardiac cycles have been reached.
    (e) Either a preset number of cardiac cycles or a preset time has been reached.

58. A technologist injects a patient with 1 mg of unlabeled stannous phosphate. After 20 min, 20 mCi of ^{99m}Tc pertechnetate is administered. This is:

    (a) An in vitro labeling procedure
    (b) An in vivo labeling procedure
    (c) A modified in vivo labeling procedure
    (d) None of the above

59. Myocardial viability can be determined by the following studies:

    (a) FDG-PET
    (b) Sestamibi after Persantine administration
    (c) ^{201}Tl without stress test
    (d) ^{201}Tl after Persantine administration
    (e) (a) and (c) only

60. Which of the following may have an effect on the time it takes to acquire a cardiac gated blood pool study?

    (a) Amount of radiopharmaceutical injected
    (b) Heart rate
    (c) Use of arrhythmia filtering
    (d) All of the above
    (e) (a) and (b) only

61. Gated cardiac blood pool scintigraphy (MUGA) is used to evaluate:

    (a) Ejection fraction
    (b) Wall motion and contraction
    (c) Myocardial ischemia
    (d) All of above
    (e) (a) and (b) only

62. For gated cardiac blood pool scintigraphy (MUGA), which of the following is not correct?

    (a) PYP needs to warm up to room temperature before reconstitution.
    (b) A good bolus injection is critical.

(c) Imaging is obtained 20 min or later after tracer injection.
(d) Both LAO and anterior views are required.

63. To perform a sestamibi myocardial perfusion study, which of the following is not correct?

    (a) NPO overnight or at least 4 h
    (b) No coffee or tea at least 12 h prior to the study for Lexiscan test
    (c) No coffee or tea at least 12 h prior to the study for Persantine test
    (d) No coffee or tea at least 12 h prior to the study for treadmill test

64. Coffee or tea is not contraindicated in which of the following studies?

    (a) Myocardial perfusion study with Persantine
    (b) Myocardial perfusion study with Lexiscan
    (c) Myocardial perfusion study with treadmill
    (d) FDG-PET study of the brain

65. To perform a sestamibi myocardial perfusion study, which of the following is correct?

    (a) Aminophylline is used to treat patient's symptoms from treadmill exercise.
    (b) Aminophylline is used to treat patient's symptoms from Persantine or dobutamine.
    (c) Aminophylline is used to treat patient's symptoms from Persantine or Lexiscan.
    (d) Aminophylline is used before the stress test to avoid any side effect.

66. To perform a pharmaceutical sestamibi myocardial perfusion study, which of the following is correct?

    (a) Sestamibi should be injected 2 min after Lexiscan administration.
    (b) Sestamibi should be injected immediately after Persantine administration.

(c) Sestamibi should be injected immediately after dobutamine administration.
(d) Sestamibi should be injected during Lexiscan administration.
(e) Sestamibi should be injected during dobutamine administration.

67. Myocardial ^{99m}Tc-PYP imaging is used for the evaluation of:

(a) Patients with suspected ischemic heart disease and chest pain
(b) Patients with suspected cardiac sarcoidosis
(c) Patients with suspected cardiac amyloidosis
(d) Myocardial viability

68. To perform a myocardial perfusion study, which of the following is correct?

(a) Rest and stress tests have to be performed on the same day.
(b) Rest imaging can be followed by stress imaging for treadmill sestamibi test.
(c) Rest imaging can be followed by stress imaging for treadmill thallium test.
(d) Rest imaging can be followed by stress imaging for Lexiscan thallium test.
(e) Rest imaging can be followed by stress imaging for Persantine thallium test.

Respiratory System Scintigraphy

9

1. In general, how many capillaries are blocked when a perfusion lung scan using ^{99m}Tc MAA is performed?

   (a) <1%
   (b) <0.1%
   (c) <0.01%
   (d) None of the above

2. Which of the following should be considered when injecting MAA?

   (a) Patient position
   (b) Blood in syringe
   (c) Filters in intravenous lines
   (d) All of the above

3. The distribution of aerosol particles in the lungs is influenced by all of the following except:

   (a) Turbulent airflow
   (b) Amount of technetium added
   (c) Rate of airflow
   (d) Particle size

4. During the equilibrium phase of a ventilation study, the patient:

   (a) Inhales O_2 and exhales ^{133}Xe
   (b) Inhales and exhales ^{133}Xe

E. Mantel et al., *Nuclear Medicine Technology*,
https://doi.org/10.1007/978-3-031-26720-8_9

(c) Inhales ^{133}Xe and exhales O_2
(d) Inhales and exhales a mixture of ^{133}Xe and O_2

5. What best describes the normal blood flow to the lungs in a supine patient?

(a) Largely uniform
(b) Decreased flow to apices relative to bases
(c) Increased flow to upper lobes
(d) None of the above

6. What best describes the normal blood flow to the lungs in an erect patient?

(a) Largely uniform
(b) Decreased flow to apices relative to bases
(c) Increased flow to upper lobes
(d) None of the above

7. During a lung perfusion study, activity is noted in the head and in the area of the kidneys. This represents:

(a) Incorrect particle size
(b) Probable metastases
(c) Free technetium in the MAA
(d) Right-to-left shunt

8. What is the method of localization for a perfusion lung study?

(a) Sequestration
(b) Compartmental containment
(c) Active transport
(d) Capillary blockade

9. Hot spots in the lungs on a perfusion study indicate:

(a) A perfusion defect
(b) Turbulent airflow
(c) Tracer injection into a supine patient
(d) Improper particle size
(e) That the blood was withdrawn into injection syringe

10. Which of the following describes the correct order in which air would pass through the respiratory anatomy?

    (a) Trachea, right bronchus, right upper lobe bronchus, bronchioles, alveoli
    (b) Trachea, left bronchus, left middle lobe bronchus, bronchioles, alveoli
    (c) Trachea, right bronchus, left lower lobe bronchus, bronchioles, alveoli
    (d) Trachea, right bronchus, right lower lobe bronchus, alveoli, bronchioles

11. Which radiopharmaceutical will be most useful in detecting delayed washout in a patient with COPD?

    (a) ^{99m}Tc MAA
    (b) ^{99m}Tc DTPA
    (c) ^{81m}Kr
    (d) ^{133}Xe
    (e) ^{99m}Tc MAG3

12. On lung ventilation images, activity is seen in the trachea and the stomach. This indicates:

    (a) A right-to-left shunt
    (b) Incorrect particle size
    (c) That the study was performed with ^{99m}Tc DTPA
    (d) That the study was performed with ^{133}Xe
    (e) A contaminated aerosol delivery system

13. Which of the following describes the correct procedure for injection of ^{99m}Tc MAA for a perfusion lung study?

    (a) Injection should be made through an existing IV line if possible.
    (b) Ensure that a minimum of 800,000 MAA particles are injected.
    (c) Small amount of blood should be withdrawn into the syringe to ensure venous access before injection.
    (d) Patient should be in the supine position.

14. Advantages of using ^{81m}Kr for ventilation lung scans include:
    (a) Ability to perfectly match perfusion and ventilation positioning.
    (b) Short half-life decreases exposure to technologists if the patient removes mask.
    (c) Ventilation and perfusion studies can be acquired simultaneously.
    (d) All of the above.
    (e) (a) and (b) only.

15. It is advisable to wait one-half hour following injection of ^{99m}Tc MAA before scanning to allow time for radiopharmaceutical clearance from the circulation.
    (a) True
    (b) False

16. Increased risk of pulmonary embolism is associated with:
    (a) Atrial fibrillation
    (b) Use of oral contraceptives
    (c) Recent surgery
    (d) All of the above
    (e) (a) and (c) only

17. The number of particles injected during a perfusion lung scan should be decreased for:
    (a) The elderly
    (b) Those with a high risk of pulmonary embolus
    (c) Patients with severe pulmonary hypertension
    (d) Asthmatics

18. Which dose would be the best choice to administer for a ventilation scan with ^{99m}Tc DTPA aerosol?
    (a) 1–2 mCi
    (b) 3–5 mCi
    (c) 8–15 mCi
    (d) 25–35 mCi

19. Lung quantitation is often used to assist physicians in the diagnosis of pulmonary embolism.

    (a) True
    (b) False

20. A patient with lung cancer is scheduled to undergo resection of the right lower lobe. Given the counts obtained by ROI for each lung segment below, what percentage of respiratory function will be lost?

Right apical	32,867
Right middle	38,952
Right lower	41,502
Left apical	39,458
Left lower	45,201

    (a) 17%
    (b) 19%
    (c) 21%
    (d) 24%
    (e) Cannot be determined from the information given

21. A perfusion lung scan shows a cold defect in the right base. If the ventilation study for the same patient is normal, what is the most probable explanation for the defect?

    (a) COPD
    (b) Pulmonary embolism
    (c) Attenuation from the heart
    (d) Pacemaker

22. If a 5,000,000-particle MAA kit is reconstituted with 70 mCi of ^{99m}Tc in 4 ml, how many particles will be in 2 ml of the reconstituted kit?

    (a) 125,000
    (b) 800,000
    (c) 1,250,000
    (d) 2,500,000

23. Regarding the reconstituted kit in Question 22, approximately how many particles will be administered to a patient who receives a 4 mCi dose?

    (a) 167,500
    (b) 575,000
    (c) 285,700
    (d) 329,000

24. A patient with pulmonary embolism can have a negative chest X-ray.

    (a) True
    (b) False

25. An MAA kit has an average of 8,000,000 particles and is reconstituted using 90 mCi of ^{99m}Tc in 5 ml at 8:00 a.m. Can the kit be used at 2 p.m. without injecting the patient with more than 800,000 particles?

    (a) Yes
    (b) No
    (c) Cannot be determined from the information given

26. Over time, the number of particles per ml in a reconstituted MAA kit will change.

    (a) True
    (b) False

27. A perfusion lung scan is being quantified. If counts obtained from an ROI around the left lung are 142,857 and the counts within the ROI about the right lung are 195,246, what percentage of perfusion is directed to the right lung?

    (a) 28%
    (b) 42%
    (c) 58%
    (d) 65%
    (e) 73%

28. The capillaries surrounding alveoli have a diameter of:

    (a) 0.3–1.0 μm
    (b) 0.4–2.0 μm
    (c) 7.0–10.0 μm
    (d) 5.0–10.0 mm
    (e) 10.0–20.0 mm

29. When ^{99m}Tc MAA breaks down after injection into a patient, what happens to the particle fragments?

    (a) They are excreted through the urine.
    (b) They are excreted by the bowel.
    (c) They are removed from the blood by the liver and spleen.
    (d) They combine with other fragments in the bloodstream and are again stopped in the lung capillary bed.

30. Ventilation images are usually performed:

    (a) With the patient supine so that perfusion is relatively uniform throughout the lungs
    (b) Anteriorly, because more emboli occur anteriorly than posteriorly
    (c) Posteriorly to reduce soft tissue attenuation and decrease distance to detector
    (d) With the patient erect, so that perfusion is relatively uniform throughout the lungs

31. The use of a nebulizer is required with which of the following radiopharmaceuticals during lung imaging?

    (a) ^{99m}Tc DTPA
    (b) ^{99m}Tc MAA
    (c) ^{127}Xe
    (d) ^{133}Xe
    (e) All of the above except (b)

32. Perfusion and ventilation lung imaging removes the need for a chest X-ray.

    (a) True
    (b) False

33. When performing a V/Q lung study with ^{99m}Tc MAA and ^{133}Xe, which portion of the exam should be performed first?
    (a) Ventilation.
    (b) Perfusion.
    (c) Either may be performed first.
    (d) They may be performed simultaneously.
34. Trapping systems for ^{133}Xe should be installed at the level of the _____, because ^{133}Xe is _____ than air.
    (a) Ceiling, heavier
    (b) Ceiling, lighter
    (c) Floor, heavier
    (d) Floor, lighter
35. Stomach visualization on a ventilation scan performed using ^{99m}Tc DTPA aerosol indicates that:
    (a) There was turbulent airflow.
    (b) The nebulizer is not functioning properly.
    (c) The patient is a smoker.
    (d) The patient swallowed some of the radiopharmaceutical.
    (e) The patient has a right-to-left cardiac shunt.
36. The radiopharmaceutical used for ventilation scanning with the highest administered dose is:
    (a) ^{133}Xe gas
    (b) ^{99m}Tc DTPA aerosol
    (c) ^{127}Xe gas
    (d) ^{81m}Kr
    (e) ^{99m}Tc MAA
37. Only half of the administered dose of ^{99m}Tc DTPA aerosol is delivered to the patient.
    (a) True
    (b) False
38. Patient education prior to a ventilation scan may:
    (a) Improve image quality
    (b) Relax the patient and thereby speed washout

(c) Reduce the radiation dose to the technologist
(d) All of the above
(e) (a) and (c) only

39. The liver may be seen in the washout phase of a ventilation study because:

(a) ^{99m}Tc DTPA aerosol was swallowed and then entered the bloodstream via the stomach.
(b) ^{133}Xe gas is fat soluble.
(c) Particle size is too small.
(d) Airflow is turbulent.

40. An advantage of ^{127}Xe over ^{133}Xe is that:

(a) It has a shorter half-life.
(b) It is inexpensive.
(c) It can be used after a ^{99m}Tc MAA perfusion study.
(d) It requires no special trapping because of the short half-life.

41. When used in conjunction with ^{99m}Tc, ^{81m}Kr can be used to perform ventilation imaging:

(a) Before
(b) During
(c) After
(d) All of the above

42. ^{133}Xe can be used to perform ventilation scans with a portable camera in the intensive care unit.

(a) True
(b) False

43. A V/Q scan can be used:

(a) To determine the likelihood of pulmonary embolism
(b) To evaluate the resolution of pulmonary embolism
(c) To quantify the differential pulmonary function before pulmonary surgery
(d) To evaluate cardiac shunts
(e) All of the above

44. When performing an aerosol ^{99m}Tc DTPA aerosol V/Q scan, which of the following is correct?

    (a) ^{99m}Tc DTPA aerosol ventilation imaging can be performed before or after perfusion imaging.
    (b) The dose used for the ventilation imaging is much higher than the perfusion imaging.
    (c) The dose delivered to the lungs of patient for the ventilation imaging is much higher than the perfusion imaging.
    (d) The count rate of the second imaging should be three to four times the count rate of the first imaging.
    (e) Both (b) and (d).
    (f) Both (c) and (d).

45. When performing a ^{133}Xe V/Q scan, which of the following is not correct?

    (a) Perfusion scintigraphy can be performed first, and if normal, ventilation scintigraphy can be omitted.
    (b) The biological half-life of the MAA in the lungs is approximately 1.5–3 h.
    (c) ^{99m}Tc MAA particles may settle in the vial with time; thus, the vial should be agitated and the syringe should be inverted prior to injection.
    (d) The number of MAA particles should be no less than 500,000.

46. When performing a ^{99m}Tc DTPA aerosol V/Q scan, which of the following is not correct?

    (a) An aerosol ventilation allows multiple projection imaging to match those obtained for perfusion imaging.
    (b) Aerosol ventilation imaging can be performed at the bedside because no special requirement for radioactive gas exhaust.
    (c) Aerosol ventilation imaging is not affected by turbulent flow in patients with COPD.
    (d) SPECT images can be obtained.

47. Advantages of V/Q scan over CTA include the following except:
    (a) Can be performed in patients with poor renal function
    (b) Can be performed in patients with pregnancy
    (c) Provides more incidental findings to explain patient's symptoms
    (d) Lower radiation exposure

48. When performing a ^{99m}Tc MAA lung perfusion scan to evaluate possible right-to-left shunting, which of the following is correct?
    (a) Multiple planar images of the chest, with stomach and kidneys included.
    (b) Multiple planar images of the chest, with thyroid included.
    (c) Multiple planar images of the chest, with both kidneys and thyroid included.
    (d) Images must include the brain.

49. For a patient with confirmed or suspected Covid-19 infection and suspected PE, which of the vollowing is correct:
    (a) Patient should have a VQ scan if there is contraindication for contrast-enhanced CT.
    (b) Patient should have a lung perfusion scan, with optional ventilation scan if patient can tolerate.
    (c) Patient should have a lung perfusion SPECT/CT, if there is contraindication for contrast-enhanced CT.
    (d) None of the above.

10 Gastrointestinal Tract Scintigraphy

1. Which of the following pairs of radiopharmaceuticals are used for simultaneously studying gastric emptying of both liquids and solids?
   (a) ^{99m}Tc sulfur colloid chicken livers and ^{111}In DTPA scrambled eggs
   (b) ^{99m}Tc albumin colloid scrambled eggs and ^{99m}Tc colloid in milk
   (c) ^{99m}Tc sulfur colloid scrambled eggs and ^{111}In DTPA in water
   (d) ^{111}In DTPA scrambled eggs and ^{99m}Tc sulfur colloid in water

2. What structure is outlined by the ROI?
   (a) Esophagus
   (b) Stomach
   (c) Common bile duct
   (d) Duodenum

E. Mantel et al., *Nuclear Medicine Technology*,
https://doi.org/10.1007/978-3-031-26720-8_10

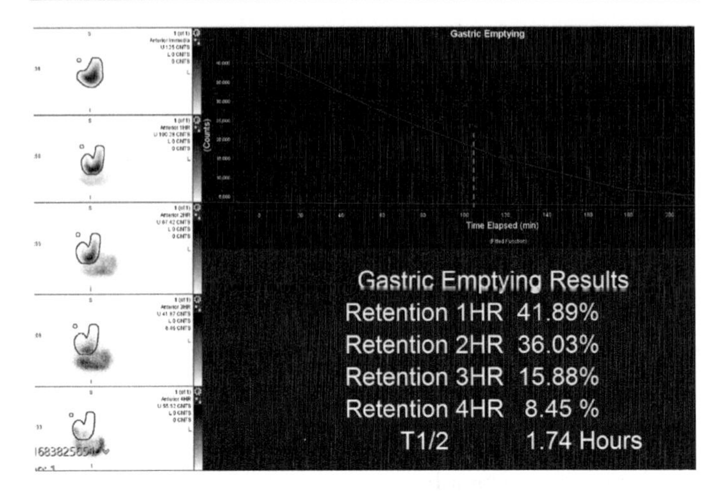

Fig. 10.1 Gastric empty scan

3. Which of the following radiopharmaceuticals can be used to study the spleen?

   (a) In vitro ^{99m}Tc-labeled autologous red blood cells
   (b) ^{99m}Tc-labeled damaged red blood cells
   (c) ^{99m}Tc sulfur colloid
   (d) All of the above
   (e) (b) and (c) only

4. How is a radiopharmaceutical administered when a LeVeen shunt is evaluated?

   (a) By intravenous injection
   (b) By intraperitoneal injection
   (c) By subcutaneous injection
   (d) By intrathecal injection

5. The ejection fraction of the gallbladder can be evaluated using:
   (a) Cimetidine
   (b) Dipyridamole
   (c) Cholecystokinin
   (d) Dobutamine

6. If no gallbladder is seen at 1 h following injection of ^{99m}Tc mebrofenin (Choletec), what should be done?
   (a) Cimetidine should be given.
   (b) Images should be taken at 24 h.
   (c) Images should be taken at 4 h.
   (d) The exam should be ended.

7. While performing a liver scan with sulfur colloid, the technologist notices that lung uptake is present. This is probably due to:
   (a) Free pertechnetate in the sulfur colloid preparation
   (b) Too much Al^{3+} ion in the pertechnetate that was used for labeling
   (c) Lung metastases
   (d) Patient being injected in the supine position

8. The function of a LeVeen shunt is to:
   (a) Treat hydrocephalus
   (b) Drain peritoneal fluid into the inferior vena cava
   (c) Drain peritoneal fluid into the superior vena cava
   (d) Divert CSF flow

9. A technologist receives a request to perform a scan with ^{99m}Tc albumin colloid to rule out acute cholecystitis. What should he or she do next?
   (a) Check if albumin colloid is available.
   (b) Alert the nuclear medicine physician to the problem.
   (c) Change the order to ^{99m}Tc IDA.
   (d) Perform the study as ordered.

10. When performing a SPECT study of the liver and spleen, the camera should be positioned:

    (a) So that the costal margin is at the bottom of the detector face
    (b) So that it touches the left side of the patient when rotating
    (c) So that it touches the right side of the patient when rotating
    (d) As close as possible but without touching the patient during rotation

11. Radiocolloids are cleared from the circulation by:

    (a) Liver parenchymal cells
    (b) Hepatocytes
    (c) Kupffer cells
    (d) Hemangiomas

12. Which radiopharmaceutical is commonly used to image cavernous hemangioma?

    (a) ^{99m}Tc albumin colloid
    (b) ^{99m}Tc RBCs
    (c) ^{99m}Tc IDA
    (d) ^{99m}Tc pertechnetate
    (e) ^{99m}Tc sulfur colloid

13. Colloid shift refers to:

    (a) Small colloid particles clumped together to form large particles which localize in the lungs
    (b) ^{99m}Tc sulfur colloid changing into albumin colloid
    (c) Increased uptake of colloid in the spleen and bone marrow relative to the liver
    (d) The redistribution of colloid within the liver over time

14. Which of the following does not involve the use of labeled RBCs?

    (a) Blood pool imaging of the liver
    (b) GI bleed imaging
    (c) Spleen imaging
    (d) Meckel's diverticulum
    (e) Both (a) and (d)

15. Cimetidine:

    (a) Prevents the gallbladder from contracting
    (b) Prevents the release of pertechnetate from gastric mucosa
    (c) Increases the uptake of pertechnetate in the gastric mucosa
    (d) Decreases peristalsis

16. The approximate time for half of the activity to empty from the stomach during a solid-phase gastric emptying exam is:

    (a) 25 min
    (b) 40 min
    (c) 90 min
    (d) 180 min

17. Esophageal reflux studies are usually performed with both solid and liquid phases.

    (a) True
    (b) False

18. Which imaging study is acquired while an abdominal binder is inflated to increasing pressures?

    (a) Meckel's diverticulum study
    (b) Gastric emptying study
    (c) Esophageal transit study
    (d) Esophageal reflux study
    (e) Gastrointestinal bleeding study

19. An appropriate adult dose and radiopharmaceutical for a Meckel's diverticulum are:

    (a) 10 mCi of ^{99m}Tc pertechnetate
    (b) 300 μCi of ^{99m}Tc sulfur colloid in water
    (c) 300 μCi of ^{99m}Tc albumin colloid in 150 mL of orange juice
    (d) 20 mCi of ^{99m}Tc-labeled red blood cells
    (e) 6 mCi of ^{99m}Tc albumin colloid

20. A patient with a bilirubin level of 35 mg/dL is scheduled for hepatobiliary imaging. The best radiopharmaceutical to use would be:

    (a) ^{99m}Tc pertechnetate
    (b) ^{99m}Tc disofenin
    (c) ^{99m}Tc mebrofenin
    (d) ^{99m}Tc sulfur colloid
    (e) ^{99m}Tc-labeled RBCs

21. If 15 min after injection of 8 mCi of ^{99m}Tc disofenin the liver is not visualized, but the heart and kidneys are, what is the most likely reason?

    (a) Too little DISIDA was injected.
    (b) The liver is not functioning properly.
    (c) The gallbladder is obstructed.
    (d) The patient is taking morphine.
    (e) The patient has a very low bilirubin level.

22. A synthetic form of cholecystokinin is:

    (a) Glucagon
    (b) Cimetidine
    (c) Pentagastrin
    (d) Sincalide
    (e) Mebrofenin

23. What are the possible effects on a hepatobiliary scan if the patient has eaten 2 h before the study?

    (a) A false positive
    (b) Nonvisualization of the gallbladder within an hour
    (c) Intermittent contraction of the gallbladder
    (d) All of the above
    (e) (a) and (b) only

24. If the maximum counts obtained from an ROI about the gallbladder are 285,000 and the minimum counts obtained from the same ROI are 187,000, what is the ejection fraction of the gallbladder?

    (a) 22%
    (b) 34%

(c) 52%
(d) 66%
(e) 73%

25. Normal gallbladder ejection fraction is:

    (a) >25%
    (b) >35%
    (c) >45%
    (d) >55%

26. What will be visualized in the first hour of a normal hepatobiliary scan?

    (a) Common bile duct
    (b) Gallbladder
    (c) Duodenum
    (d) All of the above
    (e) (a) and (b) only

27. Cholecystokinin is a hormone secreted by the duodenum that stimulates gallbladder contraction.

    (a) True
    (b) False

28. Morphine given during a hepatobiliary scan can:

    (a) Constrict the sphincter of Oddi
    (b) Enhance gallbladder filling
    (c) Shorten the study time
    (d) All of the above
    (e) (b) and (c) only

29. Sulfur colloid is best for gastrointestinal bleeding studies if:

    (a) Bleeding is intermittent.
    (b) Bleeding is active.
    (c) The bleeding site is in the right upper quadrant.
    (d) Delayed images will be planned.
    (e) None of the above.

30. Which of the following would not be useful in further examining a suspicious area of activity when evaluating a patient for lower GI bleed with ^{99m}Tc-labeled red blood cells?
    (a) Delayed imaging to visualize a change in configuration of the activity
    (b) Delayed imaging to visualize increasing activity
    (c) Use of cine mode
    (d) Anterior obliques
    (e) None of the above

31. Potassium perchlorate should not be administered to pediatric patients undergoing scanning for Meckel's diverticulum with ^{99m}Tc pertechnetate.
    (a) True
    (b) False

32. Which is the method of choice for labeling red blood cells with ^{99m}Tc if the goal is to have the least amount of free pertechnetate in the resulting dose?
    (a) In vitro.
    (b) In vivo.
    (c) Modified in vivo.
    (d) The amount of free pertechnetate will be the same in all cases.

33. Why is the labeling efficiency important for imaging of GI bleeding with ^{99m}Tc RBCs?
    (a) A small decrease in labeling efficiency may lead to a false-negative result.
    (b) If the patient is actively bleeding, the radiopharmaceutical must be prepared as quickly as possible.
    (c) ^{99m}Tc pertechnetate will be taken up by the Kupffer cells.
    (d) Free pertechnetate is secreted by the stomach and the kidneys.

34. When performing a gastric emptying study, it is important to scan the patient in the erect position to promote emptying.
    (a) True
    (b) False

35. Which of the following studies does not require any fasting prior to the examination?

    (a) Gastric emptying study
    (b) Hepatobiliary scan
    (c) Esophageal transit
    (d) Gastroesophageal reflux study
    (e) Gastrointestinal bleeding scan

36. Symptoms of cholecystitis may include:

    (a) Pain in the right upper quadrant
    (b) Back pain
    (c) Nausea
    (d) All of the above
    (e) (a) and (c) only

37. In order to reach the duodenum, bile must pass through the gallbladder.

    (a) True
    (b) False

38. Which of the following is correct?

    (a) Gastric emptying study should be terminated if the patient vomits after eating.
    (b) Gastric emptying study can be terminated if less than 10% of the meal remains in the stomach.
    (c) Both (a) and (b).

39. Which of the following can be used for gastrointestinal transit study?

    (a) ^{99m}Tc sulfur colloid
    (b) ^{99m}Tc-DTPA
    (c) ^{111}In-DTPA
    (d) All of the above
    (e) Both (a) and (c)

40. When performing GI bleeding scan using ^{99m}Tc-labeled RBC, which of the following is correct?

    (a) Dynamic images are obtained at 1 min/frame for 60 min.
    (b) If positive findings of active bleeding are observed by the technologist, he or she can stop the imaging acquisition at that time point.
    (c) If positive findings of active bleeding are observed by the technologist, he or she can stop the imaging acquisition and immediately restart a second part of imaging so that the first part of the imaging can be reviewed by a physician.
    (d) If the test is negative in the 60-min period of imaging, the patient should be sent back to the floor, but delayed imaging can be performed later if there is evidence of active bleeding later.
    (e) Both (a) and (c).
    (f) Both (a) and (d).

41. Poor ^{99m}Tc RBC labeling may be caused by:

    (a) Heparin
    (b) Iodinated contrast
    (c) Doxorubicin
    (d) Lidocaine
    (e) All of the above

42. When performing Meckel's scan to detect ectopic gastric mucosa, patients can be pretreated with the following medicines to increase the sensitivity:

    (a) Cimetidine
    (b) Pentagastrin
    (c) Glucagon
    (d) All of the above

43. To prepare a patient for hepatobiliary scintigraphy, which of the following is correct?

    (a) Patients need to be NPO for at least 4 h.
    (b) No caffeine for 12 h before the test.

(c) No opioid narcotics for at least 4 h or longer.
(d) (a) and (c) only.

44. For Meckel scintigraphy, which of the following is not correct?

 (a) Ectopic gastric mucosa in a Meckel's diverticulum is a common cause of gastrointestinal bleeding.
 (b) Meckel's diverticulum is a congenital malformation located in the distal ileum.
 (c) Meckel scintigraphy should be performed only when the patient has active bleeding.
 (d) The purpose of Meckel scintigraphy is to localize ectopic gastric mucosa in a Meckel's diverticulum.

45. For HIDA scan to evaluate biliary atresia, which of the following is not correct?

 (a) It is most commonly performed in newborn infants.
 (b) Biliary atresia is characterized by obliteration of the extrahepatic biliary system.
 (c) If no bowel activity is visualized, 4-h delayed imaging is required to complete the study.
 (d) If no bowel activity is visualized, 24-h delayed imaging is required to complete the study.

46. Patient preparation for gastric emptying or small-bowel transit study includes which of the following?

 (a) Patient should be questioned about food allergies, especially to eggs or gluten.
 (b) Patient should fast overnight or minimally for 8 h before the beginning of the procedure.
 (c) Patient should discontinue medications that affect gastrointestinal motility.
 (d) All of the above.

47. Scintigraphic small-bowel transit is a commonly performed procedure with a standardized protocol.

 (a) True
 (b) False

11 Genitourinary System Scintigraphy

1. What is the imaging protocol that was most probably used to obtain the images in Fig. 11.1?

   (a) Consecutive 1-s images for 15 s
   (b) Consecutive 3-s images for 45 s
   (c) 15 consecutive 10-s images
   (d) 15 consecutive 15-s images

2. Which of the following describes the activity on delayed static images in a patient with testicular torsion?

   (a) Decreased
   (b) Normal
   (c) Increased
   (d) Variable

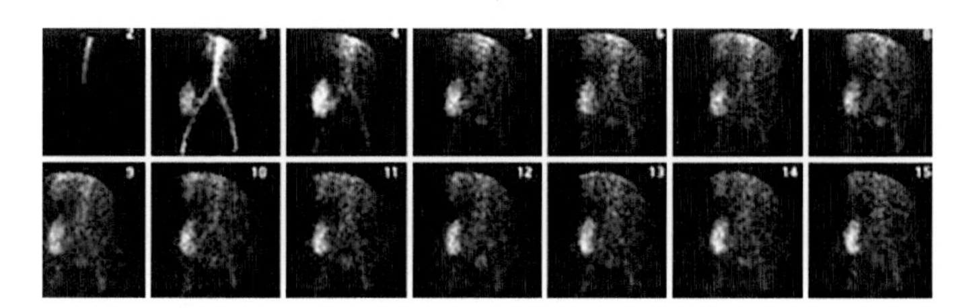

Fig. 11.1 ^{99m}Tc MAG3 renal scan, anterior view

E. Mantel et al., *Nuclear Medicine Technology*,
https://doi.org/10.1007/978-3-031-26720-8_11

3. Clearance of ^{99m}Tc MAG3 is by:

   (a) Active transport
   (b) Tubular secretion
   (c) Glomerular filtration
   (d) None of the above

4. Which renal imaging agent allows imaging at 6 h after injection?

   (a) ^{99m}Tc MAG3
   (b) ^{99m}Tc DTPA
   (c) ^{99m}Tc GH
   (d) ^{99m}Tc DMSA

5. A patient with suspected pheochromocytoma is evaluated by:

   (a) ^{123}I MIBG
   (b) ^{68}Ga DOTATATE
   (c) ^{99m}Tc MAG3
   (d) Both (a) and (b)

6. A patient who receives an injection of ^{131}I MIBG for the detection of pheochromocytoma should also receive:

   (a) Beta-blockers
   (b) Dipyridamole
   (c) Lugol's solution
   (d) Cimetidine

7. Pinhole and SPECT images may be obtained for:

   (a) ^{99m}Tc DTPA
   (b) ^{99m}Tc MAG3
   (c) ^{99m}Tc DMSA
   (d) (a) and (b)

8. ^{99m}Tc DMSA delivers a relatively high radiation dose to the kidneys because:

   (a) 20 mCi is normally injected for a renal study.
   (b) There is a long effective half-life in the kidneys.
   (c) There is a long physical half-life.
   (d) A high-energy photon is emitted.

9. Which of the following is excreted by glomerular filtration?

   (a) ^{99m}Tc DTPA
   (b) ^{99m}Tc DMSA
   (c) ^{99m}Tc MAG3
   (d) ^{99m}Tc GH

10. Indications for renal scanning with radionuclides include:

    (a) Assessing blood flow to transplanted kidneys
    (b) Assessing hydronephrosis
    (c) Assessing function of native kidneys
    (d) All of the above

11. Normal glomerular filtration rate is:

    (a) 25 mL/min
    (b) 50 mL/min
    (c) 100 mL/min
    (d) 125 mL/min

12. For visualizing intraparenchymal lesions in the kidneys, the radiopharmaceutical of choice from the list below is:

    (a) ^{99m}Tc DMSA
    (b) ^{99m}Tc DTPA
    (c) ^{99m}Tc MAG3
    (d) ^{131}I OIH

13. What is the preparation for renal functional imaging with ^{99m}Tc MAG3?

    (a) NPO for at least 4 h.
    (b) Patient must be well hydrated.
    (c) Patient must not void for at least 1 h before imaging.
    (d) Patients should not be catheterized.

14. Thirty minutes after injection of 8 mCi of ^{99m}Tc MAG3, there is significant activity remaining in the renal pelvis. What will most likely follow?

    (a) Patient will be asked to void before reimaging.
    (b) A diuretic will be administered.
    (c) Imaging will be extended for 20 min.
    (d) All of the above.

15. If after administration of furosemide a patient still has activity in the renal calyces, there is probably:

    (a) Poor renal function
    (b) Renal artery occlusion
    (c) Renal infarction
    (d) Collecting system obstruction

16. Performing an evaluation of a renal transplant includes the following considerations:

    (a) The best imaging will be obtained posteriorly.
    (b) Detector face should be centered over the right or left iliac fossa.
    (c) A lower dose of ^{99m}Tc MAG3 should be given to protect the transplant.
    (d) Uptake of radiopharmaceutical is usually delayed in the transplanted kidneys relative to native.
    (e) All of the above.

17. The left kidney is usually lower than the right because of the spleen.

    (a) True
    (b) False

18. Diuretic renography is used to evaluate:

    (a) Function of a transplant kidney
    (b) Renal artery occlusion
    (c) Split renal function
    (d) Collecting system obstruction

19. Torsion of the testicle:

    (a) Is synonymous with epididymitis
    (b) Is not painful
    (c) Is a surgical emergency
    (d) May be treated with antibiotics

20. The adult dose of ^{99m}Tc pertechnetate for imaging the testicles is:

    (a) 1–2 mCi
    (b) 2–4 mCi
    (c) 5–10 mCi
    (d) 10–20 mCi

21. The kidneys are normally perfused via the iliac artery.

    (a) True
    (b) False

22. Radionuclide cystography is most often performed to:

    (a) Determine transplant function
    (b) Evaluate renal perfusion
    (c) Detect vesicoureteral reflux
    (d) Visualize space-occupying lesions in the bladder

23. The expected bladder capacity for a 6-year-old child is:

    (a) 50 mL
    (b) 120 mL
    (c) 240 mL
    (d) 500 mL

24. The reason for calculation of expected bladder capacity before radionuclide cystography is:

    (a) To use the result in calculations of residual volume
    (b) To use the result in calculations of reflux volume
    (c) To have an idea of when maximum bladder filling will be reached
    (d) All of the above

25. A technologist is performing radionuclide cystography on a 4-year-old girl. During bladder filling with ^{99m}Tc in saline, there is leakage from around the catheter. What impact will this have?

    (a) Quantitative information will be unreliable.
    (b) The detector may become contaminated if it has not been properly protected with plastic-backed, absorbent paper.
    (c) The chance to image any reflux is lost.
    (d) (a) and (b).
    (e) (b) and (c).

26. Which of the following describes normal images from radionuclide cystography during bladder filling?

    (a) Increasing activity in the bladder over time
    (b) Homogeneous activity in both ureters
    (c) Decreasing activity in the kidneys over time
    (d) None of the above

27. Preparation for a pediatric radionuclide cystography includes:

    (a) Patient catheterization
    (b) Calculation of expected bladder capacity
    (c) Taking measures to prevent contamination of equipment, personnel, and imaging room
    (d) Emptying the patient's bladder
    (e) All of the above

28. In normal patients, a volume of up to 1.5 mL may reflux into the ureters and kidneys at maximum bladder filling.

    (a) True
    (b) False

29. Voiding images taken during radionuclide cystography should be:

    (a) Dynamic images
    (b) Acquired based on information density
    (c) 5-min images
    (d) Taken for 500 K counts

30. Which of the following is not true regarding glomerular filtration rate?

    (a) It usually becomes abnormal before serum creatinine levels become abnormal.
    (b) It is typically obtained through the use of ^{99m}Tc DTPA.
    (c) It can be determined only by taking blood and urine specimens.
    (d) It is a measure of the ability of the kidneys to clear inulin from the plasma.
    (e) All except (b).

31. Which radiopharmaceuticals can be used to determine effective renal plasma flow?

    (a) ^{99m}Tc MAG3
    (b) ^{131}I OIH
    (c) ^{99m}Tc DTPA
    (d) (a) and (b) only
    (e) (a) and (c) only

32. An advantage of GFR and ERPF measurements over other indicators of renal function such as BUN and creatinine is that the function of each kidney can be determined separately.

    (a) True
    (b) False

33. Which saline bottle would be sufficient to fill the bladder of an 8-year-old during radionuclide cystography?

    (a) 125 mL
    (b) 250 mL
    (c) 400 mL
    (d) None of the above

34. Which renal imaging agent requires the highest administered dose?

    (a) ^{99m}Tc DTPA
    (b) ^{131}I OIH
    (c) ^{99m}Tc MAG3
    (d) ^{99m}Tc DMSA
    (e) (a) and (c)

35. A technologist prepares a radiopharmaceutical for renal imaging at 8:00 a.m. The patient arrives late at 12:15 p.m., and therefore a new kit has to be made. The radiopharmaceutical being used is:

    (a) ^{99m}Tc MAG3
    (b) ^{99m}Tc DMSA
    (c) ^{99m}Tc DTPA
    (d) ^{99m}Tc GH

36. On a normal renal scan, gallbladder activity is noted. This indicates that the scan was performed using:

    (a) ^{99m}Tc DMSA
    (b) ^{99m}Tc DTPA
    (c) ^{99m}Tc GH
    (d) ^{99m}Tc MAG3

37. Dynamic imaging is typically obtained for:

    (a) ^{99m}Tc DTPA
    (b) ^{99m}Tc MAG3
    (c) ^{99m}Tc DMSA
    (d) (a) and (b)

38. The adrenal glands are:

    (a) Superior to the kidney
    (b) Inferior to the kidney
    (c) Anterior to the kidney
    (d) Posterior to the kidney

39. Which of the following is a tuft of capillaries?

    (a) Loop of Henle
    (b) Glomerulus
    (c) Renal pyramid
    (d) Collecting tubule

40. What percentage of cardiac output is directed to the kidneys?

    (a) 10%
    (b) 25%
    (c) 40%
    (d) 50%

41. On otherwise normal static images of the kidneys, the lower poles of the kidneys appear slightly decreased in intensity relative to the upper poles. Why is this?

    (a) There is probably an obstruction of the collecting system.
    (b) There are bilateral space-occupying lesions in the lower poles.
    (c) The adrenal glands are attenuating the activity from the lower poles.
    (d) The lower poles of the kidneys are situated slightly anterior to the upper poles.

42. In a patient with renal artery stenosis:

    (a) A post-captopril study will show increased GFR.
    (b) A post-captopril study will show decreased GFR.
    (c) A post-captopril study will show GFR to be unchanged.

43. When performing captopril renography, captopril should be administered:

    (a) 1 h prior to the injection of radiopharmaceutical
    (b) 20 min after the injection of radiopharmaceutical
    (c) Only if activity persists in the renal pelvis 20 min after radiopharmaceutical administration
    (d) At the same time as the injection as of radiopharmaceutical

44. Preparation for captopril renography includes:

    (a) ACE inhibitors stopped
    (b) Fasting to enhance absorption of oral captopril
    (c) Well-hydrated patient
    (d) All of the above
    (e) None of the above

45. Captopril is used when:

    (a) The patient is suspected of having a ureteropelvic obstruction.
    (b) The patient is suspected of having renovascular hypertension.
    (c) The patient has high blood pressure.
    (d) The patient is taking diuretics.

46. What will probably be done next, given the renogram shown in Fig. 11.2?

    (a) The patient will undergo captopril renography.
    (b) Posterior obliques should be taken.
    (c) Delayed images should be taken.
    (d) The exam will be ended.
    (e) None of the above.

47. When performing ACEI renography, all of the following is correct except that:

    (a) Patient should be well hydrated.
    (b) Severe hypotension may occur due to administration of ACEI.
    (c) Blood pressure and pulse should be monitored continuously throughout the study.
    (d) If a patient is already on ACEI therapy, no further ACEI is needed before ACEI renography.

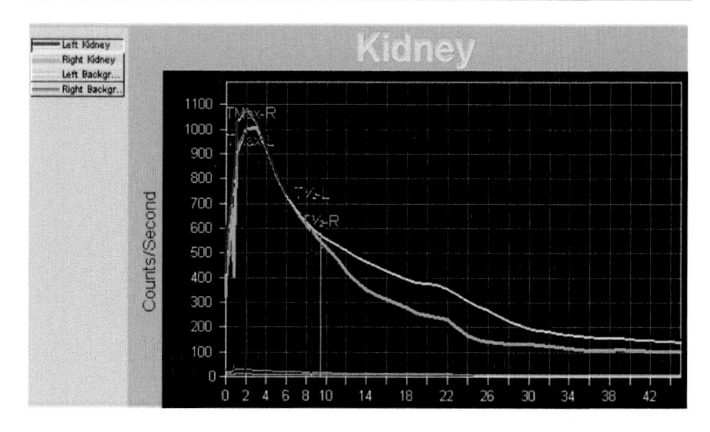

Fig. 11.2 ^{99m}Tc MAG3 renogram with lasix

Oncologic Scintigraphy 12

1. Tracer with increased uptake in tumor tissues includes:

    (a) ^{99m}Tc sestamibi
    (b) ^{201}Tl chloride
    (c) ^{67}Ga citrate
    (d) All of the above

2. What is the optimal scanning time for neoplasm when using ^{67}Ga citrate?

    (a) 4 h
    (b) 6 h
    (c) 24 h
    (d) 48 h

3. Lymphoscintigraphy is used to:

    (a) Identify metastatic lymph nodes
    (b) Identify nonmetastatic lymph nodes
    (c) Identify the sentinel node
    (d) All of the above

4. ^{111}In pentetreotide is a:

    (a) Potassium analog
    (b) Monoclonal antibody
    (c) Somatostatin analog
    (d) Radiocolloid

E. Mantel et al., *Nuclear Medicine Technology*,
https://doi.org/10.1007/978-3-031-26720-8_12

5. A HAMA response occurs because:

   (a) A kit contains pyrogens.
   (b) Monoclonal antibodies are produced from mouse cells which the human body recognizes as a foreign protein.
   (c) A patient is allergic to ^{111}In.
   (d) Monoclonal antibodies are produced from human cells that trigger an immune response in the patient.

6. Lymphoscintigraphy is often performed by _____ injection of a radiocolloid to identify the sentinel node(s).

   (a) Intradermal
   (b) Subdermal
   (c) Subcutaneous
   (d) Intratumoral

7. Sentinel node lymphoscintigraphy should be performed in a breast cancer patient with:

   (a) Known bone metastasis
   (b) Known bone metastasis
   (c) Palpable axillary nodes
   (d) No suspected metastasis

8. Sentinel node lymphoscintigraphy should include which of the following images in a patient with melanoma in the left upper back?

   (a) Images including left axilla
   (b) Images including left and right axillae
   (c) Images including bilateral axillae and groins
   (d) Either (b) or (c)

9. ^{99m}Tc sestamibi is commonly used in the evaluation of parathyroid tumor.

   (a) True
   (b) False

10. A monoclonal antibody is developed which displays cross-reaction. This means:

   (a) The antibody triggers an immune response in the patient.
   (b) The antibody will bind antigens other than the one it was formed with.

(c) The antibody can be labeled with either ^{99}mTc or ^{111}In.
(d) None of the above.

11. Examples of neuroendocrine tumor include:

(a) Pituitary adenoma
(b) Small-cell lung cancer
(c) Neuroblastomas
(d) All of the above
(e) (a) and (c) only

12. Visualization of kidneys at 48 h is _____ on a scan using ^{111}In pentetreotide and _____ on a ^{67}Ga citrate scan.

(a) Normal, abnormal
(b) Abnormal, normal
(c) Normal, normal
(d) Abnormal, abnormal

13. ^{18}F FDG can be used to image tumors because the glycolytic rate is higher in tumor than in normal tissues.

(a) True
(b) False

14. Which of the following are true regarding ^{18}F FDG tumor imaging?

(a) The patient should fast before the exam.
(b) ^{18}F FDG-PET/CT is not recommended for prostate cancer.
(c) ^{18}F FDG-PET/CT is recommended for lung cancer staging and treatment response evaluation.
(d) All of the above.

15. On a Monday morning, a technologist received two orders for an inpatient: an order of solid gastric emptying study and an order of ^{18}F FDG-PET/CT. Which of the following is correct?

(a) GI bleeding study and ^{18}F FDG-PET/CT cannot be performed on the same day.
(b) If GI bleeding study and ^{18}F FDG-PET/CT are performed on the same day, GI bleeding study should be performed first.

(c) If GI bleeding study and ^{18}F FDG-PET/CT are performed on the same day, ^{18}F FDG-PET/CT should be performed first.
(d) The order placed earlier should be performed first.

16. On a Monday morning, a technologist received two orders for an inpatient: an order of solid gastric emptying study and an order of ^{18}F FDG-PET/CT. Which of the following is correct?

    (a) Solid gastric emptying study and ^{18}F FDG-PET/CT cannot be performed on the same day.
    (b) If solid gastric emptying study and ^{18}F FDG-PET/CT are performed on the same day, gastric emptying study should be performed first.
    (c) If solid gastric emptying study and ^{18}F FDG-PET/CT are performed on the same day, ^{18}F FDG-PET/CT should be performed first.
    (d) None of the above.

17. The use of ^{18}F FDG-PET imaging in oncology takes advantage of the _____ differences between normal and neoplastic tissue.

    (a) Structural
    (b) Metabolic
    (c) Density
    (d) Hormonal

18. Hodgkin's disease is a type of:

    (a) Lung cancer
    (b) Lymphoma
    (c) AIDS
    (d) Lupus

19. F-18 fluciclovine is an FDA-approved imaging agent for what type of cancer?

    (a) Thyroid cancer
    (b) Lung cancer
    (c) Liver cancer
    (d) Prostate cancer

20. What is the half-life of Ga-68 DOTATATE?

    (a) 68 min
    (b) 78 h
    (c) 6 h
    (d) 110 min

21. Recommended dose for imaging with Ga-68 DOTATATE?

    (a) 10 mCi
    (b) 15 mCi
    (c) 20 mCi
    (d) 0.054 mCi/kg up to 5.4 mCi

22. Gallium-68 is ___________ produced.

    (a) Generator
    (b) Cyclotron

23. Gallium-68 DOTATATE is indicated for imaging what type of cancers utilizing PET?

    (a) Non-small cell lung cancer
    (b) Neuroendocrine tumors
    (c) Prostate cancer
    (d) Breast cancer

24. What is recommended dose for imaging with F-18 fluciclovine?

    (a) 10 mCi (370 MBq)
    (b) 5 mCi (185 MBq)
    (c) 20 mCi (740 MBq)

25. Which of the following is not a recommended PET tracer used to evaluate prostate cancer?

    (a) 18F-fluciclovine
    (b) 18F-FDG
    (c) 18F-NaF
    (d) 18F-piflufolastat (Pylarify)

26. Which of the following is incorrect?

    (a) Prostate-specific membrane antigen (PSMA) is a glutamate carboxypeptidase.
    (b) PSMA is a transmembrane glycoprotein.

(c) PSMA is overexpressed in prostate cancer but not expressed in normal prostate tissue.
(d) PSMA is expressed in non-prostatic malignancies including renal cell carcinoma, hepatocellular carcinoma, thyroid cancers, and gliomas.

27. PET/CT images are obtained approximately 60 min after tracer injection, except for:

    (a) 18F-fluciclovine (Axumin)
    (b) 18F-NaF
    (c) 18F-FDG
    (d) 18F-piflufolastat (Pylarify)

28. 18F-fluoroestradiol PET/CT is mainly used for:

    (a) Staging of breast cancer
    (b) Tumor aggressiveness of breast cancer
    (c) Treatment response of breast cancer
    (d) Heterogeneity of estrogen receptor (ER) expression in breast cancer

13 Infection Scintigraphy

1. Which of the following is true of ^{67}Ga citrate- and ^{111}In-labeled leukocytes?

   (a) Both require the use of a medium-energy collimator.
   (b) Both can be used to effectively image neoplasms and infections.
   (c) Both have three gamma peaks available for imaging.
   (d) All of the above.
   (e) (a) and (b) only.

2. In general, the best radiopharmaceutical to use for a suspected abdominal abscess is:

   (a) ^{67}Ga citrate
   (b) ^{99m}Tc-labeled leukocytes
   (c) ^{111}In-labeled leukocytes
   (d) ^{111}In octreotide

3. For imaging of inflammation or infection, leukocytes may be labeled with:

   (a) ^{111}In oxine
   (b) ^{99m}Tc exametazime
   (c) ^{67}Ga citrate
   (d) All of the above
   (e) (a) and (b) only

E. Mantel et al., *Nuclear Medicine Technology*,
https://doi.org/10.1007/978-3-031-26720-8_13

4. The images in Fig. 13.1 were obtained 24 h following the injection of radiopharmaceutical to visualize infection in the right leg. The radiopharmaceutical used was:

   (a) ^{99m}Tc sulfur colloid
   (b) ^{111}In-labeled WBCs
   (c) ^{67}Ga citrate
   (d) ^{99m}Tc-labeled WBCs

5. If platelets are inadvertently labeled along with leukocytes during labeling with ^{111}In oxine for imaging infection, the resulting scan may be:

   (a) False negative
   (b) False positive

6. The collection of whole blood for leukocyte labeling should be performed:

   (a) With a small-bore needle
   (b) Through an existing IV line
   (c) Using a heparinized syringe
   (d) With a shielded syringe

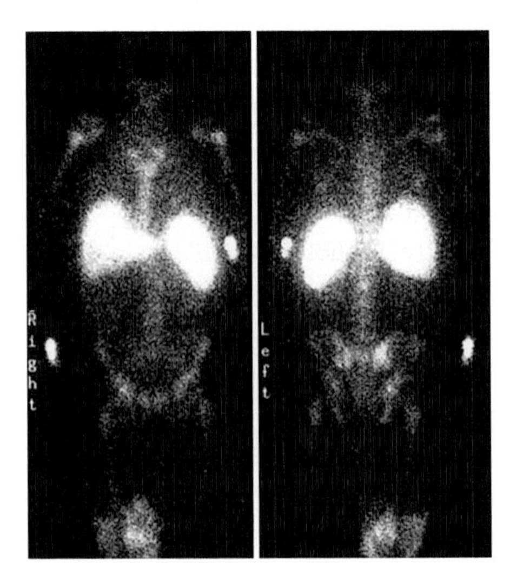

Fig. 13.1 Images obtained 24 hours post tracer injection

7. To screen for infection in a severely leukopenic patient, the best choice of radiopharmaceutical would be:

   (a) ^{99m}Tc-labeled WBCs
   (b) ^{67}Ga citrate or FDG-PET/CT
   (c) ^{111}In-labeled WBCs
   (d) ^{99m}Tc sulfur colloid

8. Gravity sedimentation involves the settling of cells:

   (a) After centrifugation
   (b) After time has passed
   (c) After the syringe or test tube is inverted
   (d) After the sample has been placed over a magnet

9. The most appropriate dose to inject for imaging inflammation with ^{111}In-labeled leukocytes is:

   (a) 0.5–1.0 mCi
   (b) 2.0–4.0 mCi
   (c) 5–10 mCi
   (d) 6–8 mCi

10. ^{67}Ga is often used with three-phase bone scans to detect:

    (a) Arthritis
    (b) Lung metastases
    (c) Osteomyelitis
    (d) Avascular necrosis
    (e) Rib fractures

11. On planar abdominal images obtained with ^{67}Ga citrate, suspicious activity is noticed. Which technique would be the first choice for further investigation of that activity?

    (a) Sequential SPECT scanning to observe change in the activity.
    (b) Subtract images obtained with ^{99m}Tc sulfur colloid.
    (c) Hydrate the patient and encourage voiding before reimaging.
    (d) Perform another exam with ^{111}In leukocytes.

12. A patient has suspected osteomyelitis of lumbar spine. Which of the following study is best indicated?
    (a) Triple-phase bone scan
    (b) ^{111}In-WBC scan
    (c) ^{99m}Tc-WBC scan
    (d) ^{67}Ga citrate scan
13. A patient has suspected abdominal infection. Which of the following study is best indicated?
    (a) Triple-phase bone scan
    (b) ^{111}In-WBC scan
    (c) ^{99m}Tc-WBC scan
    (d) ^{67}Ga citrate scan
14. To evaluate suspected infection in pediatric patients, ____ is preferred because of ____.
    (a) ^{111}In-WBC scan, better image quality
    (b) ^{111}In-WBC scan, lower dose
    (c) ^{99m}Tc-WBC scan, better image quality
    (d) ^{99m}Tc-WBC scan, lower radiation
15. When ^{111}In-WBC scan is used to evaluate osteomyelitis, ^{99m}Tc sulfur colloid imaging is commonly included because ^{99m}Tc sulfur colloid imaging may:
    (a) Increase the sensitivity of ^{111}In-WBC scan
    (b) Increase the specificity of ^{111}In-WBC scan
    (c) Decrease nonspecific uptake of ^{111}In-WBC imaging
    (d) Increase specific uptake of ^{111}In-WBC imaging
16. The differences between ^{111}In-WBC scan and ^{99m}Tc-WBC scan include:
    (a) Different half-life
    (b) Different excretion
    (c) Different imaging time
    (d) Different mechanism in detecting infection
    (e) (a)–(c)

17. The highest physiologic activity on ^{111}In-WBC scan is noted at:

    (a) Liver
    (b) Spleen
    (c) Bone marrow
    (d) GI tract
    (e) (a)–(c)

18. ^{99m}Tc sulfur colloid is cleared by the:

    (a) Liver
    (b) Spleen
    (c) Bone marrow
    (d) GI tract
    (e) (a)–(c)

19. For a patient with acute osteomyelitis, three-phase bone scan will show:

    (a) Increased uptake in the first phase
    (b) Increased uptake in the second phase
    (c) Increased uptake in the third phase
    (d) Three-phase bone scan does not directly evaluate acute osteomyelitis
    (e) (a)–(c)

20. A negative three-phase bone scan:

    (a) Is not helpful for the differential diagnosis of osteomyelitis
    (b) Is helpful for the differential diagnosis of osteomyelitis only when ^{111}In-WBC scan is included
    (c) Is helpful for the differential diagnosis of osteomyelitis only when ^{67}Ga citrate scan is included
    (d) Is more helpful for the differential diagnosis of osteomyelitis than a positive three-phase bone scan does

21. A patient had left-hip replacement and now has a three-phase bone scan to evaluate for possible loosening/infection.

    (a) Increased MDP uptake in the hip is abnormal if it is more than 6 months after surgery.
    (b) Increased MDP uptake in the hip is abnormal if it is more than 12 months after surgery.
    (c) Increased MDP uptake in the hip is abnormal if it is more than 2 years after surgery.
    (d) Absence of increased MDP uptake in the hip rules out prosthesis loosening/infection.

22. Infection and tumor are two major causes of fever of unknown origin (FUO). Patient with fever of unknown origin is recommended to be evaluated with:

    (a) FDG-PET/CT
    (b) ^{111}In-WBC scan
    (c) ^{99m}Tc-WBC scan
    (d) ^{67}Ga citrate scan

23. For evaluation of infection and inflammation, 18F-FDG-PET/CT can be used in the following conditions (choose one best answer):

    (a) Sarcoidosis
    (b) Osteomyelitis
    (c) Fever of unknown origin
    (d) Giant-cell arteritis
    (e) All of the above

24. For evaluation of infection and inflammation by 18F-FDG-PET/CT, which of the following is not correct:

    (a) The patient should avoid strenuous physical exercise within 24 h before injection.
    (b) Hyperglycemia is not an absolute contraindication for performing the study.
    (c) 18F-FDG dose adjustment (less dose) is not necessary for patients with renal failure.
    (d) Noncaloric beverages (such as water or coffee) are allowed, and NPO of at least 4 h before 18F-FDG imaging is not required for evaluating infection.

25. Normal areas of uptake for ^{67}Ga citrate include all of the following except:

    (a) Intestinal mucosa
    (b) Parathyroid glands
    (c) Liver
    (d) Epiphyses in pediatric patients

26. If a low-energy collimator is used for imaging ^{67}Ga, what will the effect be?

    (a) There is increased septal penetration
    (b) Poor spatial resolution
    (c) Decreased sensitivity
    (d) (a) and (c)
    (e) (a) and (b)

Thyroid, Parathyroid, and Salivary Gland Scintigraphy

14

1. A technologist is performing a thyroid uptake with ^{123}I sodium iodide. The capsule is counted before being administered to the patient, and 850,192 cpm is obtained. Six hours after swallowing the capsule, the counts from the patient's neck are 116,239 and from the thigh 34,982. Background for the uptake probe is 239 cpm, and the 6-h decay factor for ^{123}I is 0.730. What is the 6-h uptake?

    (a) 7.0%
    (b) 9.5%
    (c) 13.1%
    (d) 18.7%

2. Is the uptake value determined in Question 1 a normal value?

    (a) Yes
    (b) No

3. What will the effect be if a technologist places the uptake probe over the proximal thigh when counting background in the patient?

    (a) The background will be falsely elevated.
    (b) The background will be falsely decreased.
    (c) The calculated uptake will decrease.
    (d) (a) and (c).
    (e) (b) and (c).

E. Mantel et al., *Nuclear Medicine Technology*,
https://doi.org/10.1007/978-3-031-26720-8_14

4. Which of the following dietary supplements will affect thyroid uptake?
   (a) Vitamin A
   (b) Vitamin B
   (c) St. John's wort
   (d) Kelp tablets
   (e) Lactobacillus
5. Which of the following will not affect thyroid uptake?
   (a) Iodinated contrast media
   (b) Propylthiouracil
   (c) Thyroid hormones
   (d) Beta-blockers
6. Thyrotropin is also known as:
   (a) TRH
   (b) TSH
   (c) T3
   (d) T4
7. Which of the following statements is not true?
   (a) TRH stimulates the release of TSH by the anterior pituitary.
   (b) Release of T3 is inhibited by elevation of T4.
   (c) T3 and T4 are both hormones which are manufactured and released by the thyroid.
   (d) TRH is synthesized in the hypothalamus.
8. Most people have four parathyroid glands.
   (a) True
   (b) False
9. The salivary glands include:
   (a) Parotid glands
   (b) Sublingual salivary glands
   (c) Submaxillary salivary glands
   (d) All of the above
   (e) (b) and (c) only

10. Which of the following are used to image the parathyroid?
    (a) ^{99m}Tc pertechnetate
    (b) ^{99m}Tc sestamibi
    (c) ^{201}Tl chloride
    (d) (a) and (b)
    (e) (b) and (c)

11. The collimator attached to a thyroid uptake probe is:
    (a) Converging
    (b) Diverging
    (c) Pinhole
    (d) Flat field
    (e) Low energy all purpose

12. Which of the following is part of the preparation for a thyroid uptake?
    (a) NPO from midnight.
    (b) Administer Lugol's solution.
    (c) Have the patient void before administration of radiopharmaceutical.
    (d) Withhold caffeine-containing beverages for 24 h prior to examination.
    (e) None of the above.

13. When performing a thyroid uptake, the technologist neglects to count the capsule before it is administered to the patient. What are the implications?
    (a) It will not be possible to calculate uptake values.
    (b) An identical capsule must be counted in a neck phantom in order to obtain uptake values.
    (c) No decay factor will be used in the calculation of uptake values.
    (d) None of the above.
    (e) (b) and (c).

14. While taking information from a patient who is scheduled for a thyroid uptake and scan with ^{123}I sodium iodide, a technologist learns that he or she has been taking Cytomel that week. What does this mean?

    (a) Nothing; thyroid uptake will not be affected by Cytomel.
    (b) Imaging may still be obtained using ^{201}Tl chloride.
    (c) Both the uptake and scan can be performed using ^{131}I.
    (d) Study should be performed as ordered, but the uptake will be inaccurate.
    (e) None of the above.

15. What is the method of localization of ^{99m}Tc pertechnetate in the thyroid?

    (a) Sequestration
    (b) Active transport
    (c) Receptor binding
    (d) Diffusion
    (e) Phagocytosis

16. Imaging of the thyroid takes place:

    (a) Approximately 20 min after injection of ^{99m}Tc pertechnetate
    (b) 6 h after administration of an ^{123}I sodium iodide capsule
    (c) 24 h after administration of an ^{123}I sodium iodide capsule
    (d) All of the above

17. Which of the following will not optimize images of the thyroid?

    (a) Use of a pinhole collimator
    (b) Having the patient avoid swallowing during image acquisition
    (c) Hyperextending the neck
    (d) Placing a radioactive marker on the xiphoid process

18. A linear area of activity in the esophagus is seen on a thyroid image taken using ^{99m}Tc pertechnetate. What does it represent?

    (a) Sublingual thyroid
    (b) Parathyroid

(c) Pertechnetate that was secreted by the salivary glands and swallowed
(d) Parotid gland

19. The use of ^{131}I for thyroid imaging:

(a) Is common if an uptake is also planned
(b) Is typically only used when scanning the whole body for metastatic thyroid disease after thyroidectomy
(c) Delivers a lower radiation dose to the thyroid than does ^{99m}Tc pertechnetate, since only μCi amounts are administered
(d) All of the above

20. A patient receives both 3 mCi of ^{201}Tl chloride and 5 mCi of ^{99m}Tc pertechnetate, and pinhole images of the neck are taken after each administration. If the pertechnetate image is subtracted from the ^{201}Tl image, the activity that remains represents:

(a) Thyroid
(b) Parathyroid
(c) Salivary glands
(d) Hypothalamus
(e) Nonfunctioning thyroid tissue

21. The salivary glands may be imaged using:

(a) 3 mCi of ^{201}Tl
(b) 5 mCi of ^{99m}Tc sestamibi
(c) 5 mCi of ^{99m}Tc pertechnetate
(d) 200 μCi of ^{123}I sodium iodide

22. Which of the following describes delayed images at 2–3 h after injection of ^{99m}Tc sestamibi?

(a) Persistent activity in the normal thyroid with complete washout of activity in the parathyroid
(b) Persistent activity in parathyroid adenomas and decreased activity in the thyroid relative to early images
(c) Persistent activity in hyperfunctioning thyroid tissue and no activity in the parathyroid or salivary glands
(d) Activity in the salivary glands and normal parathyroid tissue

23. A hot nodule on a thyroid image will most likely be benign.
    (a) True
    (b) False

24. Which of the following are symptoms of hyperthyroidism?
    (a) Exophthalmos
    (b) Bradycardia
    (c) Cold intolerance
    (d) All of the above
    (e) (b) and (c)

25. The highest doses of therapeutic ^{131}I are given to patients with:
    (a) Grave's disease
    (b) Toxic multinodular goiter
    (c) Thyroid cancer
    (d) Chronic thyroiditis

26. A patient with a 45% uptake of ^{123}I sodium iodide at 24 h is:
    (a) Euthyroid
    (b) Hyperthyroid
    (c) Hypothyroid
    (d) Athyroid

27. The part of the thyroid that lies anterior to the trachea and is often not seen on thyroid imaging is the:
    (a) Right lobe
    (b) Left lobe
    (c) Isthmus
    (d) Parathyroid
    (e) Superior thyroid notch

28. On anterior-view thyroid images taken using ^{123}I sodium iodide, the right lobe appears to be larger than the left. The explanation for this is:
    (a) A right hemigoiter.
    (b) A hypofunctioning left lobe.
    (c) A normal finding.

(d) The patient's head was turned slightly to the left.
(e) The patient's head was turned slightly to the right.

29. Iodine is needed for the thyroid gland to synthesize T3 and T4.

    (a) True
    (b) False

30. About 90% of the thyroid hormone secreted into the blood is in the form of:

    (a) Thyroxine
    (b) Triiodothyronine
    (c) Thyroglobulin
    (d) Thyrotropin
    (e) Iodotyrosine

31. A patient with hyperparathyroidism will have:

    (a) Myxedema
    (b) Exophthalmos
    (c) Cold nodules in the thyroid
    (d) Increased fracture risk
    (e) More than four parathyroid glands

32. Why is a 6-h thyroid uptake obtained?

    (a) In case the patient does not show up the following day for the 24-h uptake
    (b) To detect hyperthyroidism when the turnover is so rapid that the 24-h uptake may be normal
    (c) In case an error is made in the calculation of the 24-h uptake
    (d) So that the patient can resume eating

33. Which patient will have the longest wait before an accurate thyroid uptake with ^{123}I can be obtained?

    (a) A patient who had a myelogram
    (b) A patient who took Synthroid
    (c) A patient who had an IVP
    (d) A patient who was given Lugol's solution
    (e) A patient who took kelp tablets

34. To stimulate secretion during salivary gland scintigraphy, which of the following is often used?

    (a) ^{99m}Tc sestamibi
    (b) Lemon juice
    (c) Lugol's solution
    (d) Captopril
    (e) Furosemide

35. Ectopic thyroid tissue may occur:

    (a) In the pelvis
    (b) In the neck
    (c) In the mediastinum
    (d) At the base of the tongue
    (e) All of the above

36. If the salivary glands are not seen on a thyroid scan obtained with the use of ^{99m}Tc pertechnetate, it may mean:

    (a) That the thyroid is hyperfunctioning and trapped the majority of the tracer
    (b) That the salivary gland function is compromised
    (c) That the salivary glands lie inferior to the thyroid and cannot be seen on anterior images
    (d) (a) and (b)
    (e) (b) and (c)

37. When would a low-energy all-purpose collimator be used for thyroid examinations?

    (a) During uptake counting
    (b) When searching for ectopic thyroid with ^{131}I sodium iodide
    (c) When performing whole body with ^{131}I sodium iodide scanning after thyroidectomy
    (d) When obtaining a blood pool image of the thyroid with ^{99m}Tc pertechnetate to differentiate cystic and solid masses
    (e) All except (a)

38. Radioactive iodine and ^{99m}Tc pertechnetate cross the placenta.

    (a) True
    (b) False

39. A technologist performs a thyroid uptake using an identical capsule to the one administered to the patient as a standard. Given the following data obtained 6 h after the capsule was swallowed, what is the uptake?

Neck	55,213 cpm
Thigh	2085 cpm
Standard	345,987 cpm
Background	48 cpm

   (a) 1.5%
   (b) 15.4%
   (c) 16.6%
   (d) 30.7%

40. An indication for a thyroid uptake is for use in calculation of the amount of radioiodine therapy for hyperthyroidism.

   (a) True
   (b) False

41. When performing a thyroid scintigraphy, Tc-99 m pertechnetate is preferred when:

   (a) Patient is not able to swallow radioiodine pill.
   (b) Patient had CT scan last week with IV contrast.
   (c) Patient is taking amiodarone.
   (d) All of the above.

42. When performing a radioiodine thyroid scintigraphy for a patient with suspected hyperthyroidism, a technologist found that the thyroid radioiodine uptake is close to zero, and there is only background activity on the images. Which of the following is correct?

   (a) The imaging time should be extended until good image of the thyroid gland is seen.
   (b) There is likely a technical problem with the gamma camera.
   (c) Patient should be re-dosed as the findings are not consistent with the provided history of hyperthyroidism.
   (d) Talk to your nuclear medicine physician.

43. When performing a radioiodine thyroid scintigraphy, the major advantage of using ^{131}I is:

    (a) Lower dose
    (b) Less radiation exposure
    (c) Better imaging quality
    (d) Higher sensitivity for detecting occult lesions

44. What is the typical dose for a parathyroid scan?

    (a) 20 mCi ^{99m}Tc-sestamibi
    (b) 30 mCi ^{99m}Tc-sestamibi
    (c) 10 mCi $^{99m}TcO_4^-$
    (d) 20 mCi $^{99m}TcO_4^-$

45. What is the route of administration for a parathyroid scan dose?

    (a) Orally
    (b) 10-min IV infusion
    (c) Intravenous bolus
    (d) Intramuscular

46. Dual-isotope parathyroid imaging uses what tracers?

    (a) ^{123}I and ^{131}I
    (b) ^{99m}Tc sestamibi and ^{123}I
    (c) $^{99m}TcO_4^-$ and ^{131}I
    (d) ^{99m}Tc sestamibi and ^{131}I

47. What is the typical dose of ^{99m}Tc pertechnetate for a thyroid scan?

    (a) 0.5 mCi
    (b) 1.0 mCi
    (c) 5.0 mCi
    (d) 20.0 mCi

48. ^{131}I dose for a diagnostic thyroid scan?

    (a) 100 uCi
    (b) 200 uCi
    (c) 500 uCi
    (d) 1 mCi

49. ^{131}I decays by which method?

    (a) Alpha decay
    (b) Beta decay
    (c) Gamma decay

50. What is the energy of ^{131}I?

    (a) 140 keV
    (b) 364 keV
    (c) 168 keV
    (d) 511 keV

51. Whole-body ^{131}I imaging is performed with what dose?

    (a) 2–5 mCi ^{131}I
    (b) 200 uCi ^{123}I
    (c) 200 uCi ^{131}I
    (d) 2–5 mCi ^{123}I

52. What is the half-life of ^{123}I?

    (a) 6 h
    (b) 13 h
    (c) 73 h
    (d) 8 days

53. Because $^{99m}TcO_4^-$ is trapped but not organified by the thyroid, imaging should begin __________ after injection.

    (a) At 10 min
    (b) Within 30 min
    (c) At 4 h postinjection
    (d) At 1 day postinjection

54. Thyroid scintigraphy is utilized to evaluate:

    (a) Thyroid function
    (b) Functional status of nodules
    (c) Size and location of thyroid tissue
    (d) All of the above

55. $^{99m}TcO_4^-$ is administered

(a) Orally
(b) Intravenously
(c) Intramuscularly

56. How is ^{123}I administered?

(a) Intravenously
(b) Intramuscularly
(c) Orally

Non-imaging Procedures and Radionuclide Therapy

15

1. Radioiodine therapy is given for:

    (a) Hyperthyroidism
    (b) Thyroid cancer
    (c) Grave's disease
    (d) All of the above
    (e) (a) and (b) only

2. What is the typical adult dose administered for an Azedra® therapy?

    (a) 100 mCi
    (b) 200 mCi
    (c) 400 mCi
    (d) 500 mCi

3. What is the patient preparation prior to a Lutathera® therapy administration?

    (a) No preparation
    (b) Discontinuation of somatostatin analogs
    (c) NPO for 6 h
    (d) None of the above

4. Route of administration of Y-90 microspheres is:

    (a) Intravenous
    (b) Intra-arterial

E. Mantel et al., *Nuclear Medicine Technology*,
https://doi.org/10.1007/978-3-031-26720-8_15

(c) Oral
(d) Intramuscular

5. Which RBC tagging method requires no blood handling?

(a) In vitro.
(b) In vivo.
(c) Modified in vivo.
(d) All methods listed above require blood handling.

6. ^{89}Sr chloride is used to treat:

(a) Malignant ascites
(b) Polycythemia vera
(c) Bone pain caused by metastases
(d) Grave's disease
(e) (a) and (c)

7. ^{89}Sr chloride should be administered:

(a) Via intracavitary injection
(b) Via direct venous injection
(c) Through a patent intravenous line
(d) Orally

8. ^{89}Sr is effectively shielded by:

(a) Lead pigs
(b) Plastic syringes
(c) Paper
(d) None of the above

9. What is the half-life of ^{223}radium?

(a) 8 days
(b) 11.4 days
(c) 6 h
(d) 73 h

10. What is the labeling efficiency using the in vivo tagging?

(a) 85%
(b) 90%
(c) 95%
(d) None of the above

11. ^{223}Radium emits multiple energies. 95.3% is emitted as __________.

    (a) Gamma rays
    (b) Beta rays
    (c) Alpha rays
    (d) X-rays
    (e) A and B
    (f) A and C
    (g) None of the above

12. Planning procedures prior to Y-90 microspheres calculate lung shunt. Lung shunt values with greater than 20% is a contraindication for the microsphere therapy.

    (a) True
    (b) False

13. Which microspheres are glass?

    (a) TheraSpheres®
    (b) SirSpheres®
    (c) Neither are glass

14. A centrifuge is needed for what method of RBC tagging?

    (a) In vivo
    (b) Modified in vivo
    (c) In vitro labeling

15. What is the typical dose regimen for Lutathera®?

    (a) 200 mCi dose every 8 weeks for a total of four doses
    (b) 200 mCi dose every 4 weeks for a total of four doses.
    (c) Single 200 mCi dose
    (d) Single 500 mCi dose

16. Mapping studies with ^{99m}Tc MAA is performed to evaluate what?

    (a) Appearance of liver to lung shunt
    (b) Nothing of significance

17. Beta emitters are effective for therapy because:

    (a) They have a short range in soft tissue.
    (b) They do not harm healthy tissue.
    (c) They can be used for imaging as well as for therapy.
    (d) They have short half-lives.

18. What is the meaning of polycythemia?

    (a) An excess of white blood cells
    (b) An excess of platelets
    (c) An excess of red blood cells
    (d) An excess of plasma

19. Which of the following measure the amount of red cells in circulating blood?

    (a) Red cell volume
    (b) Hemoglobin
    (c) Hematocrit
    (d) (a) and (c)
    (e) All of the above

20. The normal life span of a red blood cell is _____ days, following which it is removed from circulation by the _____.

    (a) 120, spleen
    (b) 60, spleen
    (c) 120, bone marrow
    (d) 60, bone marrow

21. Red blood cells are also known as:

    (a) Leukocytes
    (b) Erythrocytes
    (c) Granulocytes
    (d) Platelets
    (e) (a) and (d)

22. An empty syringe is used to draw 5 mL of blood. After several minutes, the blood appears separated into a liquid and a solid portion. The liquid portion is called:

    (a) Serum
    (b) Plasma

(c) Anticoagulant
(d) None of the above

23. The functions of white blood cells include:

(a) Ingestion of bacteria by phagocytosis
(b) Antibody production
(c) Production of cellular immunity
(d) All of the above
(e) (b) and (c) only

24. A patient should be supine for at least 15 min before the start of a plasma volume determination because the plasma volume _____ when a patient is standing.

(a) Increases
(b) Decreases

25. If the dose rate measured at the bedside of a radioiodine therapy patient is 10 mrem/h, how long may a visitor (who is seated at bedside) stay?

(a) 8 min
(b) 12 min
(c) 20 min
(d) 33 min

26. A patient receives 45 mCi of ^{131}I to treat thyroid cancer. A survey taken at 3 m immediately after administration of the dose reveals 60 mrem/h. If a survey taken 12 h later shows a dose rate of 30 mrem/h, how many mCi is left in the patient?

(a) 15.5 mCi
(b) 22.5 mCi
(c) 35 mCi
(d) 40 mCi

27. A patient advocate decides to question a patient undergoing radioiodine therapy regarding his or her opinion of the care provided. The dose at 1.5 m from the patient is 5 mrem/h, and it will take at least 30 min for him or her to interview the patient. What should be done?

(a) Have him or her sit 3 m from the patient.
(b) Have him or her sit at 1.5 m from the patient, and let a colleague take over questioning after 20 min.
(c) Have him or her interview the patient the following day.
(d) Any of the above.

28. According to the Code of Federal Regulations, at what point may a patient who has undergone radioiodine therapy leave the hospital?

(a) When another person is unlikely to receive a total effective dose of 0.5 rem or 5 mSv from exposure to the patient
(b) If the remaining dose in the thyroid is less than 33 mCi
(c) If the remaining dose in the thyroid is less than 35 mCi
(d) If the time spent in the hospital is greater than four times the physical half-life of the isotope
(e) (a) and (b)

29. Which of the following objects in a patient's room have the potential to become contaminated from a patient who has received radioiodine?

(a) Television
(b) Mattress
(c) Bathroom fixtures
(d) All of the above
(e) (b) and (c) only

30. Preparations for radioiodine therapy include:

(a) NPO from midnight
(b) Screening for pregnancy or breastfeeding
(c) Discontinuance of antithyroid drugs for 5–7 days prior to treatment
(d) All of the above

31. Which of the following are used to calculate the dose of ^{131}I needed for administration in a patient with hyperthyroidism?

(a) Weight of the gland
(b) Percent uptake
(c) Presence or absence of nodules
(d) All of the above

32. A patient has a 6-h thyroid uptake of 25% and receives a 20 μCi dose of ^{131}I for a thyroid uptake prior to radioiodine therapy. What will the concentration of activity in the patient's thyroid gland be at 6 h if his thyroid weighs 45 g?

    (a) 11 mCi/g
    (b) 11 μCi/g
    (c) 0.11 μCi/g
    (d) 1.1 μCi/g

33. Which of the following make ^{131}I suitable for therapy?

    (a) Uptake in thyroid tissue regardless of function
    (b) Alpha emissions
    (c) Short half-life
    (d) Beta emissions

34. Alternatives to radioiodine therapy for patients with hyperthyroidism are:

    (a) Surgery
    (b) Antithyroid drugs
    (c) Thyroid storm
    (d) All of the above
    (e) (a) and (b) only

35. The wash method of labeling red blood cells for the determination of red cell volume involves repeated centrifugation of the collected sample. The advantage of this technique is that:

    (a) Anticoagulants are unnecessary.
    (b) Free chromate ion is removed.
    (c) The use of expensive ascorbic acid is avoided.
    (d) Less ^{51}Cr can be used.

36. If a technologist mistakenly administers ^{32}P chromic phosphate intravenously, what is likely to be the result?

    (a) The urine will be bluish green for about a week.
    (b) Severe radiation damage to the liver.
    (c) Hypothyroidism will occur within 1 year.
    (d) None of the above.

37. Discharge instructions for a patient who underwent therapy with ^{131}I may include all of the following except:

    (a) Patient is encouraged to increase fluids and void frequently.
    (b) Patient should avoid close contact with others.
    (c) Patient should collect excreta and store for ten half-lives.
    (d) Patient should flush toilet at least twice after use.

38. A survey taken at 1 m from a patient who received 35 mCi of ^{131}I revealed a dose rate of 9 mrem/h. If the survey made 24 h later shows 6.9 mrem/h, what is the remaining dose in the patient, and can he or she be discharged?

    (a) 27 mCi, yes
    (b) 27 mCi, no
    (c) 4.3 mCi, yes
    (d) 4.3 mCi, no
    (e) Cannot be determined from the information given

39. The NRC requires ^{131}I therapy patients to have private rooms, although they may share with another ^{131}I therapy patient.

    (a) True
    (b) False

40. According to the NRC, records of dose rate measurements taken immediately after administration of radioiodine to therapy patients must be kept for:

    (a) 6 months
    (b) 1 year
    (c) 3 years
    (d) 5 years

41. After injection of ^{125}I human serum albumin for determination of plasma volume, it is important to withdraw a small amount of blood into the syringe and reinject it, to ensure that the entire radioactive dose is administered.

    (a) True
    (b) False

42. A patient who receives ^{89}Sr chloride for palliation of pain from bone metastases does not have to be admitted to the hospital.

    (a) True
    (b) False

43. Which of the following is used in the treatment of non-Hodgkin's lymphoma?

    (a) ^{82}Rb
    (b) ^{32}P chromic phosphate
    (c) ^{90}Y ibritumomab tiuxetan
    (d) ^{153}Sm microspheres

44. Patients who have previously demonstrated allergic reactions to mouse proteins have an increased risk of allergic response to:

    (a) ^{131}I tositumomab
    (b) ^{90}Y ibritumomab tiuxetan
    (c) ^{153}Sm lexidronam
    (d) (a) and (b) only
    (e) (b) and (c) only

45. To perform a thyroid scintigraphy, which of the following is correct?

    (a) ^{99m}Tc pertechnetate has the advantage of no special patient preparation.
    (b) ^{99m}Tc pertechnetate has the advantage of high dose and high count.
    (c) ^{99m}Tc pertechnetate has the advantage of lower radiation exposure.
    (d) All of the above.

46. To perform a thyroid scintigraphy, which of the following is correct?

    (a) If ^{99m}Tc pertechnetate is used, imaging can be obtained at 24 h after tracer injection.

(b) If ^{99m}Tc pertechnetate is used, imaging can be obtained after 2 h but before 24 h after tracer injection.
(c) If ^{99m}Tc pertechnetate is used, thyroid uptake can be obtained either 2 or 24 h after tracer injection.
(d) If ^{99m}Tc pertechnetate is used, thyroid uptake can be obtained 10 min after tracer injection.

47. To perform a whole-body radioiodine scintigraphy for thyroid cancer, which of the following is not correct?

(a) Patient should be on low-iodine food for 5–7 days.
(b) Patient should discontinue thyroid hormone replacement or have received Thyrogen injection before the study.
(c) Pregnancy is an absolute contraindication if ^{131}I is used.
(d) Breastfeeding should be stopped and cannot be resumed for either ^{123}I or ^{131}I test.

48. What is the half-life of radium-225 dichloride?

(a) 11.4 days
(b) 6 h
(c) 78 h
(d) 110 min

49. Radium-225 dichloride is currently FDA approved for patients with what disease process?

(a) Metastatic lung cancer
(b) Metastatic thyroid cancer
(c) Metastatic prostate cancer

50. How is radium-225 dichloride administered?

(a) Intravenous injection
(b) Intrathecal injection
(c) Orally
(d) Inhalation

51. What is the recommended dose of ^{223}radium dichloride to be administered?

(a) 1.49 μCi/kg of body weight
(b) 5 mCi

(c) 0.56 μCi/kg of body weight
(d) 1.35 mCi/kg of body weight

52. What is an indication for an Azedra® therapy?

(a) Thyroid cancer
(b) Pheochromocytoma
(c) Hyperthyroidism
(d) None of the above

53. What is an anticoagulant?

(a) Morphine
(b) Heparin
(c) Acetazolamide
(d) Lugol's solution

54. What is the half-life of ^{131}I iobenguane?

(a) 6 days
(b) 8.1 days
(c) 73 h
(d) 6 h

55. What is the method of administration of Azedra®?

(a) Orally
(b) Intravenously
(c) Intra-arterially
(d) Intrathecally

56. An anticoagulant is not used in which RBC labeling method?

(a) In vitro
(b) Modified in vivo
(c) In vivo

57. What is the half-life of Lu-177?

(a) 6.6 days
(b) 6.6 h
(c) 8 days
(d) 73 h

58. What is the typical dose regimen for Xofigo®?

    (a) 6 injections every 4 weeks
    (b) 4 injection every 6 weeks
    (c) Single 100 μCi dose
    (d) Single 200 μCi dose

59. What is the patient preparation prior to Azedra® therapy?

    (a) No patient prep
    (b) NPO × 4 h
    (c) Administration of thyroid-blocking agent

60. What is the method of administration of Lutathera®?

    (a) IV bolus
    (b) IV infusion over 30–40 min
    (c) Intramuscularly
    (d) Orally

Patient Care

16

1. Which of the following concerning an IV drip should a technologist monitor while a patient is under his or her care?
   - (a) Height of the infused substance
   - (b) Kinks in tubing
   - (c) Pain and swelling
   - (d) All of the above
2. Ambulatory means:
   - (a) Emergent
   - (b) Able to walk
   - (c) Bedridden
   - (d) In a wheelchair
3. Which of the following is not a method for decreasing the spread of HIV?
   - (a) Using a protective gown and gloves
   - (b) Using disposable needles only once
   - (c) Obtaining a detailed sexual history
   - (d) Using a protective face mask
4. If bleeding occurs during withdrawal of a needle following injection, one should:
   - (a) Apply a tourniquet
   - (b) Apply pressure

E. Mantel et al., *Nuclear Medicine Technology*,
https://doi.org/10.1007/978-3-031-26720-8_16

(c) Alert a physician
(d) Apply ice

5. NPO means:

(a) No preparation for exam
(b) Nothing by mouth
(c) Patient may drink water but should not eat
(d) None of the above

6. Which of the following should be considered when using patient restraint devices?

(a) Restricted circulation
(b) Attenuation
(c) Comfort
(d) All of the above
(e) (a) and (c) only

7. A patient who is aphasic cannot:

(a) Walk
(b) Breathe while lying flat
(c) Sit up
(d) Talk

8. If several patients receiving a dose of radiopharmaceutical from the same vial experience an adverse reaction, this is likely a(*n*):

(a) Anaphylactic reaction
(b) Allergic reaction
(c) Pyrogenic reaction
(d) Radiation sickness

9. Which of the following tests does not require the patient to be NPO for some time period prior to scanning?

(a) Hepatobiliary imaging
(b) Gastrointestinal bleeding study
(c) Exercise–redistribution myocardial scan with ^{210}Tl
(d) Schilling test

10. Of the following types of disease transmission, for which does the Centers for Disease Control recommend wearing a biosafety mask?
    (a) Airborne
    (b) Contact
    (c) Droplet
    (d) (a) and (c) only
    (e) All of the above

11. The standard precautions recommended by the Centers for Disease Control are designed for use:
    (a) With patients presenting a risk for contact transmission
    (b) With all patients who present in the healthcare setting
    (c) With patients presenting a risk for airborne transmission
    (d) With patients presenting a risk for droplet transmission

12. A nosocomial infection is:
    (a) In the nasal passage
    (b) Transmitted by coughing or sneezing
    (c) Acquired while in the hospital
    (d) Always less serious in nature

13. Before any procedure has begun on a patient, his or her identity should be checked:
    (a) According to the accompanying chart/medical record
    (b) Verbally with the patient if possible
    (c) On the wristband
    (d) By all of the above methods

14. If a patient is having seizures, the technologist should:
    (a) Try to restrict the patients' movements
    (b) Try to grasp the patient's tongue
    (c) Start CPR
    (d) Clear the area around the patient to minimize the risk of injury
    (e) All of the above

15. Infectious waste disposal involves:

    (a) Leakproof containers
    (b) Puncture-resistant containers
    (c) The universal symbol for biohazard
    (d) All of the above
    (e) (a) and (b) only

16. Gloves should be worn anytime there is contact with:

    (a) Blood and body fluids
    (b) Broken skin
    (c) Mucous membranes
    (d) All of the above
    (e) (a) and (b) only

17. According to the standard precautions issued by the Centers for Disease Control, when should gloves be changed?

    (a) Between procedures on two different patients
    (b) Between procedures on the same patient
    (c) Every half hour
    (d) All of the above
    (e) (a) and (b) only
    (f) (a) and (c) only

18. According to the Centers for Disease Control, handwashing should be performed:

    (a) Before contact with a patient
    (b) After contact with a patient
    (c) After contact with a patient if gloves were worn
    (d) All of the above
    (e) (a) and (b) only

19. Which of the following statements is not true regarding sharp instruments?

    (a) Reusable instruments should be cleaned, disinfected, and sterilized before reuse.
    (b) Nonreusable sharp instruments should be recapped before disposal into a biohazard container.

(c) Single-use needles should be disposed using puncture-resistant containers.
(d) None of the above.

20. A nursing mother who undergoes a lung perfusion exam utilizing 4 mCi of ^{99m}Tc MAA should:

(a) Continue nursing as usual.
(b) Interrupt breastfeeding for 12.6 h.
(c) Interrupt breastfeeding for 36 h.
(d) Discontinue breastfeeding.

21. Patients who receive an injection of an investigational radiopharmaceutical must:

(a) Have their urine collected and stored until it decays to background
(b) Sign a patient-informed consent form
(c) Have made a living will
(d) Be inpatients

22. Which of the following is true concerning radioactivity and pregnancy?

(a) Radioiodine will cross the placenta.
(b) Hydration and frequent voiding will reduce the fetal absorbed dose.
(c) The NRC total dose allowed to an embryo/fetus from occupational exposure is 0.5 rem.
(d) All of the above.

Positron-Emission Tomography

17

1. Which of the following radionuclides is not a positron emitter?
   (a) ^{82}Rb
   (b) ^{15}O
   (c) ^{18}F
   (d) ^{14}C
   (e) ^{68}Ga
2. When a neutron-deficient nucleus emits a positron:
   (a) Atomic mass number decreases by 1.
   (b) Atomic mass number increases by 1.
   (c) Atomic number increases by 1.
   (d) Atomic mass number is unchanged.
3. Which of the following has the shortest half-life?
   (a) ^{82}Rb
   (b) ^{13}N
   (c) ^{15}O
   (d) ^{18}F
4. Which of the following will not affect the distribution of ^{18}F-FDG on a PET image?
   (a) Intense physical activity the day before imaging.
   (b) Serum insulin level.
   (c) Serum glucose level.

E. Mantel et al., *Nuclear Medicine Technology*,
https://doi.org/10.1007/978-3-031-26720-8_17

(d) Bladder catheterization.
(e) All of the above affect distribution.

5. Normal ^{18}F-FDG distribution would show the least activity in the:

(a) Brain
(b) Bone
(c) Bladder
(d) Myocardium

6. All positron-emitting isotopes are produced in a cyclotron.

(a) True
(b) False

7. ^{18}F-FDG-PET is not an important tool for:

(a) Restaging of colorectal cancer
(b) Monitoring response to treatment of non-Hodgkin's lymphoma
(c) Detecting *Helicobacter pylori*
(d) Imaging metastases in breast cancer

8. Which of the following is true regarding ^{18}F-FDG?

(a) Its distribution in the brain is related only to blood flow.
(b) It is a potassium analog.
(c) It has a half-life of 60 min.
(d) It is taken up by disease-free myocardium.

9. ^{18}F-FDG-PET images show a map of the _____ distribution in the body.

(a) Oxygen
(b) Insulin
(c) Glucose
(d) Potassium

10. What length of time should a lactating female who undergoes ^{18}F-FDG-PET scanning wait before resuming breastfeeding?

(a) 4 h
(b) 24 h

(c) 1 week
(d) Breastfeeding should be discontinued

11. Because of the relatively high energy of the photons detected in PET imaging, attenuation correction does not need to be performed.

(a) True
(b) False

12. What part of a PET scanner quality control regimen is necessary for the computation of attenuation factors?

(a) Coincidence timing calibration
(b) Normalization correction
(c) PMT gain adjustment
(d) Blank scan

13. Which of the following describes a random coincidence?

(a) The simultaneous detection of annihilation photons originating from a single positron
(b) The simultaneous detection of annihilation photons originating from a single positron although one of the photons has been scattered before reaching the detector
(c) The simultaneous detection of photons originating from different positrons

14. Which of the following is *not* a scintillation crystal found in PET cameras:

(a) Lutetium oxyorthosilicate (LSO)
(b) Bismuth germanate (BGO)
(c) Lead sulfate ($PbSO_4$)
(d) Gadolinium oxyorthosilicate (GSO)

15. PET images may be reconstructed using:

(a) Coincidence detection
(b) Filtered back projection
(c) K-space filling
(d) Block detection

16. Following a positron annihilation, two 511 keV photons are emitted in opposite directions. If one of the photons is deflected in the body:

    (a) Only the unscattered photon will contribute to the image.
    (b) The deflected photon may be detected outside the correct LOR.
    (c) The LOR may contain an angle.
    (d) The deflected photon will have a higher energy than the undeflected photon.

17. A patient has undergone myocardial PET scans using ^{13}N ammonia and ^{18}F-FDG. There is little inferoapical activity on the ammonia scan, but there is homogeneous uptake in the same area on the FDG images. This suggests that:

    (a) An incorrect dose of ^{13}N has been administered.
    (b) The inferoapical area may revascularize.
    (c) There is little viable tissue in the inferoapical wall.
    (d) The system requires a quality control check.

18. Patient A and Patient B are injected with the same dose of ^{13}N ammonia for a myocardial PET scan. The patients are of the same height, but Patient A weighs 120 lbs. while Patient B weighs 180 lbs. From which patient would more true coincidences be detected?

    (a) Patient A
    (b) Patient B

19. Chewing gum between injection of ^{18}F-FDG and PET scanning will have no effect on the resulting images.

    (a) True
    (b) False

20. Why is it important to know if a patient has an inflammatory condition before performing an ^{18}F-FDG-PET scan?

    (a) ^{18}F-FDG and certain anti-inflammatory medications cause severe reactions when used simultaneously.
    (b) ^{18}F-FDG can exacerbate inflammatory conditions.

(c) Some inflammatory conditions cause areas of increased uptake of ^{18}F-FDG.

21. Patient preparation for an ^{18}FDG-PET scan of the whole body includes:

    (a) NPO the day before imaging
    (b) Fasting state of at least 4–6 h prior to injection
    (c) Encouraging intense physical activity the day before imaging

22. ^{18}F-FDG is useful for imaging tumors because:

    (a) It is a glucose analog.
    (b) There is an elevated rate of glycolysis in tumors.
    (c) It becomes trapped in tumor cells.
    (d) All of the above.

23. Which dose and route of administration are the best choices when performing a whole-body PET scan with ^{18}F-FDG and a dedicated PET scanner?

    (a) 2–4 mCi IV
    (b) 2–4 mCi IM
    (c) 10–20 mCi IV
    (d) 10–20 mCi IM
    (e) 30–35 mCi IV

24. Fill in the blank. High blood glucose levels may _____ ^{18}F-FDG uptake in tumors.

    (a) Decrease
    (b) Increase

25. Relative to one another, ^{82}Rb and ^{82}Sr are:

    (a) Isobars
    (b) Isotopes
    (c) Isotones
    (d) None of the above

26. What determines the lower limit of spatial resolution in PET scanners?

(a) Timing window
(b) Detector type
(c) Positron range
(d) Compton scatter

27. Parenteral nutrition should be discontinued for several hours prior to ^{18}FDG imaging.

(a) True
(b) False

28. Two unrelated annihilation photons are detected within the timing window and judged to be in coincidence. This is called a:

(a) True coincidence
(b) Random coincidence
(c) Scatter coincidence

29. Noncolinearity refers to:

(a) Annihilation photons that are not emitted at exactly 180° to one another due to particle momentum
(b) The detection of two annihilation photons, at least one of which has been redirected by Compton scatter
(c) The distance a positron travels before undergoing annihilation

30. On an ^{18}F-FDG-PET scan of the whole body, muscle uptake can be decreased by:

(a) Administering diazepam before injection of ^{18}F-FDG
(b) Maintaining the patient in the supine position during uptake
(c) Minimizing chewing, talking, and swallowing during uptake
(d) All of the above
(e) (a) and (b) only

31. Dedicated PET scanners use:

(a) Flat-field collimation
(b) Electronic collimation

(c) Converging collimation
(d) No collimation

32. Standardized uptake values are useful for:

(a) Differential diagnosis of tumor type
(b) Assessing tumor size changes
(c) Assessing response to treatment

33. A patient presents for PET scanning with ^{18}FDG for suspected tumor of the left axilla. Which injection site should be avoided?

(a) Right antecubital vein
(b) Left antecubital vein
(c) Right foot
(d) Left foot

34. The recommended dose of gallium-68 DOTATATE (Netspot®) is:

(a) 0.054 mCi/kg, up to 5.4 mCi
(b) 10 mCi
(c) 1 mCi
(d) 25 mCi

35. F-18 PSMA is a PET imaging agent for patients with ________.

(a) Breast cancer
(b) Lung cancer
(c) Prostate cancer

36. Indications for performing Netspot® PET imaging are:

(a) Localization of somatostatin receptor-positive neuroendocrine tumors
(b) Prostate cancer
(c) Breast cancer
(d) Lung cancer

37. Axumin (fluciclovine F-18) is a PET imaging tracer used to image:
    (a) Breast cancer
    (b) Prostate cancer
    (c) Lung cancer
    (d) Infection

38. What is the administration method for F-18 PSMA?
    (a) Intravenous
    (b) Intramuscular
    (c) Oral
    (d) Inhalation

39. Imaging with fluciclovine F-18 (Axumin) should be initiated:
    (a) 3–5 min postinjection
    (b) 30 min postinjection
    (c) 60 min postinjection
    (d) 90–120 min postinjection

40. Netspot® can be used to image both pediatric and adult patients.
    (a) True
    (b) False

41. Typical dose of F-18 PSMA is:
    (a) 15 mCi
    (b) 9 mCi
    (c) 12 mCi
    (d) 5 mCi

42. What is the delay time between injection and scanning with F-18 PSMA?
    (a) Immediate postinjection
    (b) 15 min postinjection
    (c) 60 min postinjection
    (d) 2 h postinjection

43. Ga-68 DOTATATE is generator produced.
    (a) True
    (b) False

44. The typical dose of fluciclovine F-18 is:
    (a) 10 mCi
    (b) 15 mCi
    (c) 20 mCi
    (d) 25 mCi

45. There are no contraindications to using Axumin (fluciclovine F-18).
    (a) True
    (b) False

46. Patient preparation includes being NPO prior to the F-18 PSMA injection.
    (a) True
    (b) False

47. Ga-68 DOTATATE PET imaging, from skull to mid-thigh, can be acquired:
    (a) Immediately postinjection
    (b) 15 min postinjection
    (c) 40–90 min postinjection
    (d) 2–4 h postinjection

48. Tumors that do not bear somatostatin receptors will not be visualized.
    (a) True
    (b) False

49. F-18 PSMA is used to scan the female patient population.
    (a) True
    (b) False

50. The half-life of Ga-68 DOTATATE is:

    (a) 68 min
    (b) 110 min
    (c) 6 h
    (d) 78.3 h

51. Patient should be encouraged to increase hydration prior to the study and to increase frequency of voiding after the administration of F-18 PSMA.

    (a) True
    (b) False

52. Ga-68 DOTATATE PET should be performed prior to administering long-acting somatostatin analogs. When can short-acting analogs be used?

    (a) Up to 24 h before
    (b) 2 weeks before
    (c) Never
    (d) Time does not matter

53. How is Ga-68 DOTATATE administered?

    (a) Oral
    (b) Intravenous
    (c) Intramuscular
    (d) Inhalation

Multimodality Imaging

18

1. Which one of the following statements is true about hybrid PET/CT or SPECT/CT imaging systems?

   (a) The same detector is used for both modalities.
   (b) Images do not have to be checked for patient motion.
   (c) Daily QC takes less time than for stand-alone systems.
   (d) The two modalities are acquired simultaneously.
   (e) The CT provides anatomic correlation for the SPECT or PET.

2. Which one of the following statements is true about the CT in SPECT/CT or PET/CT?

   (a) CT is slower than previous methods of performing attenuation correction.
   (b) CT images must be scaled and smoothed before being used for attenuation correction.
   (c) CT has worse spatial resolution than SPECT or PET.
   (d) Truncation of the body in the CT does not affect the SPECT or PET images.
   (e) Surgical or dental implants made from high-Z materials do not affect attenuation correction.

E. Mantel et al., *Nuclear Medicine Technology*,
https://doi.org/10.1007/978-3-031-26720-8_18

3. Which of the following is *not* a potential source of artifacts in PET/CT or SPECT/CT imaging?

   (a) CT beam hardening
   (b) Anatomic coverage
   (c) CT contrast medium
   (d) Respiratory motion
   (e) Metal implants

4. Which of the following is not a barrier to combining an MRI system with either a SPECT or a PET system?

   (a) The duration of MRI scans makes sequential scanning lengthy.
   (b) PMTs do not function in high magnetic fields.
   (c) MRI scanners are expensive.
   (d) MRI scanners do not produce ionizing radiation.

5. If performing a CT for attenuation correction only, why would the CT parameters be significantly altered from those for a diagnostic CT scan?

   (a) To save time.
   (b) To reduce the radiation exposure to the patient.
   (c) To increase patient throughput.
   (d) You would not alter the parameters.
   (e) None of the above.

6. A CT scan to define the area of the body to be scanned is called a:

   (a) Blank scan
   (b) Quick scan
   (c) Scout scan
   (d) Topogram
   (e) None of the above
   (f) Both (c) and (d)

7. CT artifacts can be caused by:

   (a) The operator
   (b) The scanner

(c) The patient
(d) All of the above
(e) None of the above

8. How often should the CT calibration be checked and a tube warm-up performed?

   (a) Daily
   (b) Weekly
   (c) Monthly
   (d) Quarterly

9. IV contrast must always be administered during a PET/CT scan.

   (a) True
   (b) False

10. Artifacts due to respiration mismatch between CT and SPECT or PET are most likely to be seen near:

    (a) Lung apices
    (b) Dome of the diaphragm/liver
    (c) Maxillary sinuses
    (d) Pulmonary artery

11. If a patient moves between the CT and the SPECT or PET, which two of the following corrections will be affected?

    (a) Attenuation
    (b) Scatter
    (c) Decay
    (d) Normalization
    (e) Dead time

12. Select two reasons that metal implants in patients create problems for SPECT/CT and PET/CT systems:

    (a) Additional weight of the patient may exceed table's load limit.
    (b) High CT numbers from metal overestimate the attenuation that photons emitted by radiopharmaceutical will experience.

(c) Artifacts in CT images due to the presence of metal may propagate to attenuation maps.
(d) Metal disturbs magnetic field of CT.

13. Refer to Fig. 18.1. In this fused PET/CT transaxial slice through the mandible, what is the likeliest explanation for the area of activity extending outside the boundary of the CT (see *arrow*)?

    (a) Facial edema
    (b) Patient tilted head between CT and PET
    (c) Benign cyst
    (d) Lupus

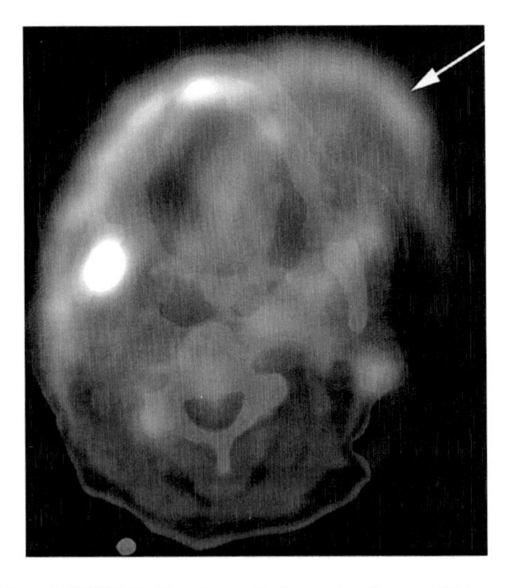

Fig. 18.1 Fused PET/CT slice through the patient's mandible

19 Appendix 1: Mock Examination

1. In which radiopharmaceutical is ^{99m}Tc left in the valence state of +7?

   (a) Sulfur colloid
   (b) Macroaggregated albumin
   (c) MAG3
   (d) Sestamibi

2. A technologist receives a request for a patient to receive ^{153}Sm for palliation of pain from bone metastases. The patient has a leukocyte count of 3500 and 40,000 platelets. What should happen next?

   (a) The order should be changed to ^{89}Sr.
   (b) The therapy should be delayed until the platelet count is at least 60,000.
   (c) The therapy should be delayed until the leukocyte count is below 2400.
   (d) The ^{153}Sm dose should be ordered.

E. Mantel et al., *Nuclear Medicine Technology*,
https://doi.org/10.1007/978-3-031-26720-8_19

3. A sulfur colloid kit requires the addition of 1–3 mL containing no more than 500 mCi of ^{99m}Tc pertechnetate. In addition, 1.5 mL of Solution A must be added before boiling, and 1.5 mL of Solution B must be added after cooling. If the kit is correctly prepared, what is the maximum concentration of ^{99m}Tc it could contain?

   (a) 83 mCi/mL
   (b) 125 mCi/mL
   (c) 167 mCi/mL
   (d) 500 mCi/mL

4. A Hine–Duley phantom is useful for testing the spatial resolution and spatial linearity with only one image.

   (a) True
   (b) False

5. The wall of the heart indicated by the arrow in Fig. 19.1 is the:

   (a) Septal
   (b) Lateral
   (c) Inferior
   (d) Anterior

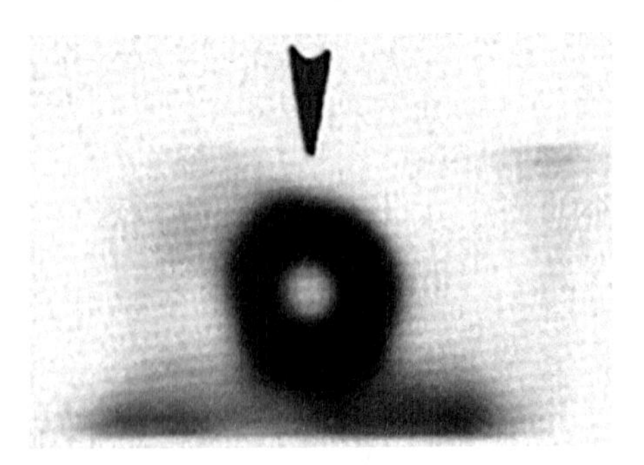

Fig. 19.1 Short axis view of myocardium

6. The advantage of a lung ventilation scan performed with ^{133}Xe is:

   (a) The ability to match perfusion and ventilation positioning
   (b) The ability to detect delayed washout
   (c) The ability to perform the exam at the patient's bedside
   (d) The ability to perform perfusion before ventilation without increasing the dose given for the ventilation study

7. A diphosphonate kit contains 180 mCi of ^{99m}Tc in 30 mL when it is prepared at 8:00 a.m. Immediately, a 20 mCi dose is withdrawn for a bone scan. If the patient arrives late at 9:30 a.m. and half the volume is accidentally discharged, how much volume from the kit must now be added to the syringe to correct the dose to 20 mCi? (No other doses have been withdrawn meanwhile, and the decay factor for 1.5 h is 0.841.)

   (a) 1.5 mL
   (b) 2.0 mL
   (c) 2.3 mL
   (d) 2.5 mL
   (e) 2.7 mL

8. A whole-body bone scan film shows a good posterior image, but the anterior view appears diffuse and with greater soft tissue activity. What is the probable cause?

   (a) Wrong collimator used for the anterior
   (b) Increased patient-to-detector distance on the anterior relative to the posterior
   (c) Low tagging efficiency of the radiopharmaceutical
   (d) Patient not sufficiently hydrated
   (e) None of the above

9. Which of the following will not reduce the probability of contamination from perspiration and saliva of a therapy patient?

   (a) Disposable eating utensils
   (b) Plastic covering over the mattress

(c) Covering the telephone receiver with plastic
(d) Collecting the patient's waste and storing until decayed to background

10. Why would an image of the brain be obtained during a lung perfusion scan using ^{99m}Tc MAA?

(a) To rule out brain metastases
(b) To detect low tagging efficiency
(c) To detect a right-to-left cardiac shunt
(d) To see if pulmonary emboli have traveled into the cerebral vessels

11. Bone marrow imaging is performed using:

(a) ^{99m}Tc MDP
(b) ^{99m}Tc HDP
(c) ^{99m}Tc sulfur colloid
(d) ^{153}Gd
(e) Both (a) and (b)

12. All of the following have an effect on the concentration of eluate obtained from a wet generator except:

(a) Time since last elution
(b) Volume of saline added to the charging port
(c) Size of collection vial
(d) Amount of ^{99}Mo activity present on the alumina column
(e) Both (b) and (c)

13. A ^{99}Mo/^{99m}Tc generator has an elution efficiency of 89%. If 1.1 Ci of ^{99m}Tc is present on the column, what is the estimated activity of eluate that will be obtained?

(a) 97.9 Ci
(b) 979 mCi
(c) 979 μCi
(d) 97.9 mCi

14. ^{99m}Tc DTPA has been used for all of the following except:

(a) Aerosol lung scanning
(b) Renal flow study

(c) Brain flow study
(d) Shunt patency
(e) Spleen scanning

15. Which of the following is not used as an anticoagulant?

(a) Heparin
(b) EDTA
(c) Ascorbic acid
(d) ACD solution

16. If the concentration of a solution desired is 2 mCi/mL and the 325 mCi of solute is contained in a volume of 2.5 mL, how much water should be added to create the solution?

(a) 160.0 mL
(b) 162.5 mL
(c) 650.0 mL
(d) 125.0 mL

17. The inferior wall of the heart is supplied by the:

(a) Left anterior descending branch
(b) Left circumflex artery
(c) Right coronary artery
(d) Left coronary artery

18. The uptake of ^{18}F-FDG by tumors is based on:

(a) A blood–brain barrier breakdown
(b) The higher glycolytic rate of tumors relative to normal tissue
(c) Active transport
(d) Antibody–antigen reactions
(c) Receptor binding

19. ^{111}In delivers a lower radiation dose to the patient than ^{67}Ga because it has only one gamma emission, whereas ^{67}Ga has four.

(a) True
(b) False

20. If dipyridamole is supplied in a 50 mL vial containing 250 mg and a patient weighs 205 lb., how much volume will need to be withdrawn from the vial to prepare his or her dose of 0.56 mg/kg?

    (a) 5.6
    (b) 10.4
    (c) 15.9
    (d) 22.6

21. If 450 mCi of ^{99m}Tc eluted from a generator contains 200 μCi of ^{99}Mo, is the eluate within the standards of radionuclidic purity set forth by the NRC?

    (a) Yes.
    (b) No.
    (c) The NRC does not regulate radionuclidic purity.
    (d) It cannot be determined from the information given.
    (e) This is an example of radiochemical impurity rather than radionuclidic impurity.

22. If SPECT images taken with a 360° rotation, 32 stops for 15 s/stop, and matrix of 64 × 64 have significant star artifact, which imaging parameter would be best to change and to correct the image?

    (a) Change the rotation to 180°.
    (b) Change the number of stops to 64.
    (c) Change the time/stop to 20 s.
    (d) Change the matrix to 128 × 128.
    (e) None of the above.

23. The images in Fig. 19.2 were taken 2 h following the intrathecal injection of a radiopharmaceutical. These are images from a:

    (a) LeVeen shunt study
    (b) Cisternogram
    (c) Esophageal transit study
    (d) CSF shunt patency study
    (e) Radionuclide venography

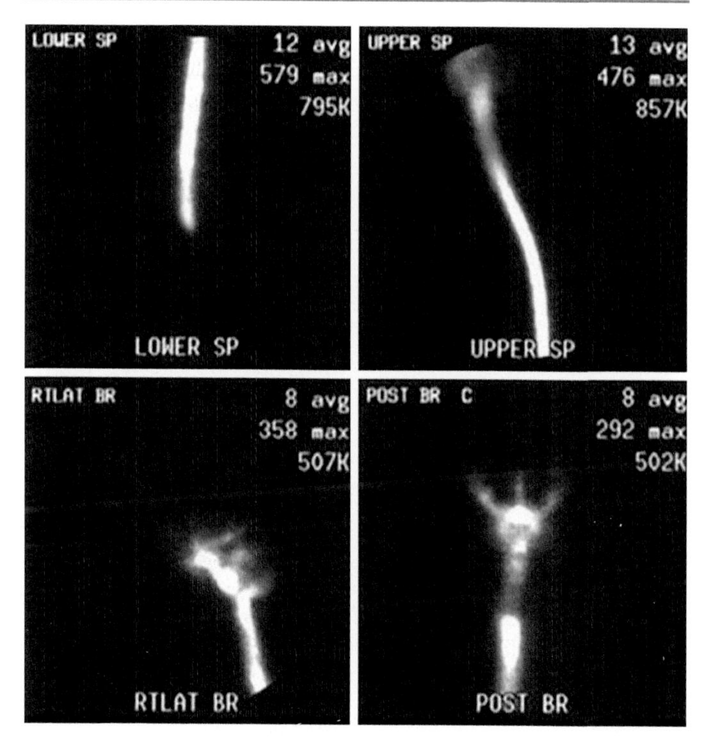

Fig. 19.2 Delay imaging, 2 hours post injection

24. Which radiopharmaceutical is best to use for renal scanning if a determination of effective renal plasma flow is also desired?
   (a) ^{99m}Tc MAG3
   (b) ^{99m}Tc GH
   (c) ^{99m}Tc DMSA
   (d) ^{99m}Tc DTPA

25. ^{68}Ga-DOTATATE PET is a radiolabeled:
   (a) Fatty acid
   (b) Glucose analog
   (c) Amino acid
   (d) Neuropeptide

26. Which of the following cannot be studied with a MUGA study?
    (a) Left ventricular ejection fraction
    (b) Regional wall perfusion
    (c) Stroke volume
    (d) Regional wall motion

27. Free ^{99m}Tc pertechnetate in a bone scan kit will result in activity in the:
    (a) Breasts
    (b) Thyroid
    (c) Gastric mucosa
    (d) Lungs
    (e) (b) and (c)

28. A superscan is the result of:
    (a) Renal insufficiency
    (b) Diffuse skeletal metastases
    (c) Chemotherapy response
    (d) Paget's disease
    (e) None of the above

29. Which of the following is not true regarding ^{99m}Tc HMPAO?
    (a) It crosses the intact blood–brain barrier.
    (b) Distribution of ^{99m}Tc HMPAO follows cerebral flow.
    (c) It is also known as exametazime.
    (d) It does not significantly redistribute in the brain.
    (e) None of the above.

30. Which of the following may cause a false-positive hepatobiliary scan?
    (a) If the patient has recently eaten
    (b) If the patient has been fasting for an extended period (>24 h)
    (c) The use of sincalide
    (d) (a) and (b)
    (e) (b) and (c)

31. Phagocytosis of ^{99m}Tc albumin colloid is performed by the:

(a) Liver parenchymal cells
(b) Kupffer cells
(c) Hepatocytes
(d) Red blood cells

32. Which will perfuse first following an intravenous injection of ^{99m}Tc DTPA?

(a) Lungs
(b) Left ventricle
(c) Kidney
(d) Liver

33. A patient with high blood pressure undergoes renography followed by captopril renography, and the curve of renal uptake from the second exam is lower than the first. What does this mean?

(a) The patient is rejecting a transplanted kidney.
(b) The patient has renovascular hypertension.
(c) The patient has an obstruction in the collecting system.
(d) The patient has high blood pressure.
(e) None of the above.

34. The left and right lobes of the thyroid are connected by the:

(a) Pyramidal lobe
(b) Major calyx
(c) Isthmus
(d) Loop of Henle

35. The exposure rate at the surface of a package to be shipped is 50 mrem/h. What label is required?

(a) DOT Radioactive White I.
(b) DOT Radioactive Yellow II.
(c) DOT Radioactive Yellow III.
(d) No radioactive label is required.

36. If the exposure rate at 5 m from a source is 75 mR/h, what will it be at 2 m?

    (a) 12 mR/h
    (b) 36 mR/h
    (c) 245 mR/h
    (d) 469 mR/h

37. If a patient has persistent activity in the renal pelvis during renography, furosemide administration may:

    (a) Assist in the diagnosis of renovascular hypertension
    (b) Help rule out mechanical obstruction
    (c) Cause a transplant to be rejected
    (d) Decrease the radiation burden on the bladder
    (e) None of the above

38. The NRC requires the use of tongs for moving vials between the shields and the dose calibrator.

    (a) True
    (b) False

39. Which of the following may be used for thyroid uptake and scanning?

    (a) ^{99m}Tc pertechnetate
    (b) ^{123}I sodium iodide
    (c) ^{201}Tl chloride
    (d) ^{99m}Tc sestamibi
    (e) All of the above

40. The lowest point on a time-activity curve generated over the cardiac cycle of a patient following a gated blood pool study represents:

    (a) End systole
    (b) End diastole
    (c) Injection point
    (d) Stroke volume
    (e) The QRS complex

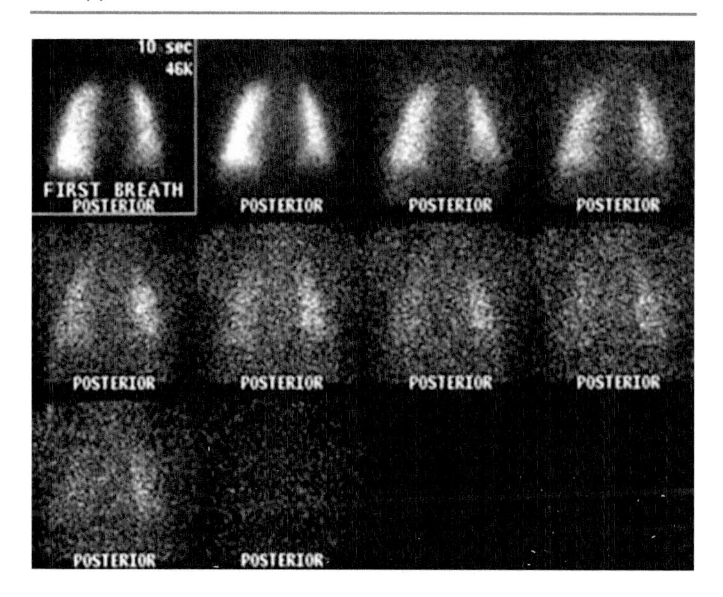

Fig. 19.3 Ventilation lung scan, anterior projection

41. The images in Fig. 19.3 were obtained during a ventilation scan taken with the use of:

    (a) ^{99m}Tc MAA
    (b) ^{133}Xe gas
    (c) ^{81m}Kr gas
    (d) ^{99m}Tc DTPA aerosol

42. Pinhole collimation is not commonly used during:

    (a) Scrotal scans
    (b) Bone scans for AVN
    (c) Thyroid imaging
    (d) V/Q scan

43. A sentinel node is:

    (a) The first draining lymph node from a primary cancer
    (b) The node containing cancer which is most distant from the primary tumor site

(c) The common iliac node
(d) Always ipsilateral to the tumor

44. Which method of data collection allows the greatest flexibility for manipulation of data after collection?

(a) Frame mode
(b) List mode
(c) There is no difference between the two methods concerning data manipulation

45. The second phase in a four-phase bone scan takes place:

(a) A few minutes after injection
(b) 2–4 h after injection
(c) 6 h after injection
(d) 24 h after injection

46. The NRC recommends but does not require the use of syringe shields during preparation and administration of radiopharmaceuticals.

(a) True
(b) False

47. Which of the following is true regarding albumin colloid?

(a) It is less expensive than sulfur colloid.
(b) The particle size is smaller than that of sulfur colloid.
(c) A lower dose is required for liver imaging.
(d) It must be boiled during preparation.

48. For a patient with intermittent gastrointestinal bleeding, the best radiopharmaceutical to locate the bleed is:

(a) ^{99m}Tc sulfur colloid
(b) ^{99m}Tc denatured RBCs
(c) ^{99m}Tc WBCs
(d) ^{99m}Tc DTPA
(e) None of the above

49. Which is not a beta emitter?

(a) ^{32}P
(b) ^{131}I

(c) ^{81m}Kr
(d) ^{89}Sr

50. Which are used for the imaging of neoplasms?

(a) ^{201}Tl chloride
(b) ^{131}I MIBG
(c) ^{99m}Tc sestamibi
(d) All of the above
(e) (a) and (b) only

51. Constancy of a dose calibrator must be performed:

(a) Daily
(b) Weekly
(c) Monthly
(d) Yearly
(e) At installation and following repair

52. The valence state of ^{99m}Tc is usually reduced when preparing kits through the use of:

(a) Nitrate
(b) Oxygen
(c) Stannous ions
(d) Alumina

53. First-pass cardiac studies require:

(a) A compact bolus of radiopharmaceutical
(b) Gating
(c) Multiple crystal imaging system
(d) Extended imaging time
(e) All of the above

54. Most of the thyroid hormone in the circulation is:

(a) T3
(b) T4
(c) TSH
(d) TRF

55. Which of the following guidelines apply to position a patient for a flow study of native kidneys?

    (a) Center the detector over the iliac fossa of interest.
    (b) Iliac crests should be centered over the detector face.
    (c) Sternal notch should be at the top of the field of view.
    (d) Xiphoid process should be over the upper part of the detector face.

56. ^{133}Xe gas may be used:

    (a) In rooms that are held at a positive pressure to surrounding areas
    (b) In patient rooms if they are private
    (c) In imaging rooms that have a ceiling vent
    (d) In rooms that are held at a negative pressure to surrounding areas
    (e) In all of the above as long as oxygen is not in use

57. Given the following data, calculate the left ventricle ejection fraction.

Net counts, end systole	20,582
Net counts, end diastole	44,653

    (a) 18.5%
    (b) 46.1%
    (c) 53.9%
    (d) 68.5%

58. Calculation of gallbladder ejection fraction requires:

    (a) Cholecystokinin
    (b) Cimetidine
    (c) Pentagastrin
    (d) ^{99m}Tc mebrofenin
    (e) (a) and (d)

59. Which study does not require any fasting before the examination?

    (a) Esophageal transit
    (b) Gastroesophageal reflux

(c) Schilling test
(d) Hepatobiliary scan
(e) Meckel's diverticulum

60. The usual adult dose for hepatobiliary imaging with ^{99m}Tc iminodiacetic acid is:

(a) 2–4 mCi
(b) 5–8mCi
(c) 10–12 mCi
(d) 15–20 mCi

61. What is the effective half-life of a radiopharmaceutical with a physical half-life of 6 h and a biological half-life of 10 h?

(a) 2.6 h
(b) 3.8 h
(c) 4.3 h
(d) 5.2 h

62. If a technologist absorbs a dose rate of 2.5 mrem/h while working 3 ft from a therapy patient, what distance will need to be maintained in order to absorb only 2 mrem/h?

(a) 2.5 ft
(b) 3.4 ft
(c) 4.0 ft
(d) 4.2 ft

63. At least 30 million counts should be obtained when assessing the uniformity of a planar imaging system.

(a) True
(b) False

64. Records of dose calibrator linearity must be kept for:

(a) 1 year
(b) 3 years
(c) 5 years
(d) 10 years
(e) Until license expiration

65. Transport index refers to:

    (a) The type of packaging required for a shipment of radioactive material
    (b) The type of labeling required for a shipment of radioactive material
    (c) The exposure rate measured at 1 m from the package
    (d) The NRC isotope group label

66. If a vial of ^{131}I contains 100 mCi on December 1, approximately how much activity will it contain on December 17 provided that none is withdrawn?

    (a) 6.25 mCi
    (b) 12.5 mCi
    (c) 25 mCi
    (d) 50 mCi

67. Which of the following is an example of radiochemical impurity?

    (a) Presence of ^{99}Mo in ^{99m}Tc eluate
    (b) Presence of free pertechnetate in ^{99m}Tc HDP
    (c) Presence of Al^{3+} in ^{99m}Tc
    (d) Presence of pyrogens in reconstituted kits

68. What does dyspnea mean?

    (a) Increased urine production
    (b) Difficulty walking
    (c) Temporary cessation of breathing
    (d) Labored breathing
    (e) Paradoxical wall motion

69. Which of the following is not used to image the heart?

    (a) ^{201}Tl chloride
    (b) ^{99m}Tc sestamibi
    (c) ^{99m}Tc diphosphonate
    (d) ^{99m}Tc RBCs

70. Arrhythmia filtering rejects cardiac cycles:

   (a) That have an R–R interval outside the preset limits
   (b) That have a too short S–T segment
   (c) That have a too high QRS complex
   (d) All of the above
   (e) (a) and (b) only

71. If a SPECT is being performed using a single-head camera involving 64 projections and 52,000 counts are collected during each 20-s stop, what are the total counts for the study and the total acquisition time?

   (a) 1 million counts, 20 min
   (b) 1.1 million counts, 21 min
   (c) 3.3 million counts, 21 min
   (d) 3.3 million counts, 25 min

72. A SPECT study is required to consist of two million counts. If there are 64 stops and the count rate is 65,000 cpm, how many seconds should each stop last?

   (a) 13 s
   (b) 21 s
   (c) 29 s
   (d) 31 s
   (e) 48 s

73. If a MDP kit contains 230 mCi of ^{99m}Tc in 9 mL at 7:00 a.m., how much will have to be withdrawn for a 20 mCi dose at 1:00 p.m.?

   (a) 0.6 mL
   (b) 0.8 mL
   (c) 1.6 mL
   (d) 1.9 mL
   (e) 2.1 mL

74. In 20 mL of ^{99m}Tc eluate, there is 30 μg of Al^{3+} present. Can this eluate be used for reconstituting kits for patient injection?

(a) Yes
(b) No

75. If 800 mCi of ^{99m}Tc is eluted from a ^{99}Mo/^{99m}Tc generator, what is the maximum amount of ^{99}Mo allowed in the eluate by the NRC?

(a) 8 μCi
(b) 12 μCi
(c) 80 μCi
(d) 120 μCi
(e) 200 μCi

76. If a technologist stands 2 ft from a radioactive source and receives 1 mrem/h from it, what will he or she absorb by stepping 1 ft closer to the source?

(a) 0.25 mrem/h
(b) 1.5 mrem/h
(c) 2.0 mrem/h
(d) 4.0 mrem/h

77. A locked storage closet next to the radiopharmacy containing phantoms and sealed sources has a measured exposure rate of 3 mR/h. It should be posted with a sign reading:

(a) Caution: Radioactive Materials.
(b) Caution: Radioactive Area.
(c) Caution: High Radiation Area.
(d) Grave Danger: Very High Radiation Area.
(e) No sign is necessary.

78. A technologist is preparing an HDP kit at 8:00 a.m. If the schedule of doses required is as follows and the eluate available has 7 mCi/mL, what is the minimum volume of ^{99m}Tc pertechnetate that should be added?

9:00	Bone scan	20 mCi
10:00	Bone scan	20 mCi
11:00	Bone scan	20 mCi

(a) 8.6 mL
(b) 10.8 mL
(c) 60.0 mL
(d) 75.9 mL

79. When injecting ^{99m}Tc MAA, which of the following are true?

(a) It is best not to withdraw blood into the syringe before injection.
(b) The patient should be supine.
(c) The radiopharmaceutical should be gently agitated before injection.
(d) All of the above.
(e) (a) and (b) only.

80. Following a spill of 50 mCi of ^{99m}Tc mebrofenin in the nuclear medicine department, the first priority is to:

(a) Notify the NRC.
(b) Notify the RSO and/or chief technologist.
(c) Perform an area survey to determine exposure rate in the area.
(d) Contain the spill by covering it and restricting access to the area.
(e) Collect glass pieces.

81. The annual TEDE limit for a technologist's hands is:

(a) 5 mrem
(b) 5 rem
(c) 15 mrem
(d) 50 mrem
(e) 50 rem

82. Center of rotation must be determined for:

(a) Each collimator used
(b) Each matrix used
(c) Each zoom factor used
(d) All of the above

83. If the HVL of lead for ^{131}I is 3 mm and a lead vial shield holding the iodine is 6 mm thick, what percentage of the original exposure rate will remain?

    (a) 75%
    (b) 50%
    (c) 25%
    (d) 12.5%

84. 1 mL of a liquid is assayed and contains 206 mCi. If it is diluted with 9 mL and subsequently is assayed at 180 mCi, what is the geometric correction factor?

    (a) 0.75
    (b) 0.87
    (c) 0.96
    (d) 1.14
    (e) 1.20

85. Increasing the matrix size of a SPECT acquisition:

    (a) Decreases the imaging time
    (b) Decreases the storage space needed
    (c) Increases spatial resolution
    (d) Increases count density
    (e) All of the above

86. An MAA kit has an average of 900,000 particles. If 50 mCi of ^{99m}Tc pertechnetate in 7 mL is added, how many particles will a patient who gets a 3 mCi dose receive?

    (a) 72,000
    (b) 129,000
    (c) 240,000
    (d) 514,000

87. To minimize oxidation of stannous chloride, which of the following may be added to radiopharmaceutical kits?

    (a) Oxygen
    (b) Water

(c) Bacteriostatic saline
(d) Nitrogen

88. On an intrinsic field uniformity image, the area of increased activity around the image is called:

(a) Halo effect
(b) Edge packing
(c) Flare phenomenon
(d) Septal penetration

89. The photomultiplier tube functions to:

(a) Convert light to electrical signal
(b) Emit light in response to a gamma ray
(c) Filter radiation not originating perpendicular to the detector face
(d) Discard signal from background radiation

90. Film badges should be worn:

(a) Between the shoulder and waist
(b) Except while doing paperwork
(c) Only during preparation and administration of radiopharmaceuticals
(d) All of the above

91. If the half-life of a daughter isotope is longer than the half-life of the parent, no equilibrium is possible:

(a) True
(b) False

92. ^{18}F-fluciclovine (Axumin®) PET/CT is used for what indication?

(a) Suspected Alzheimer's disease
(b) Lymphoma
(c) Suspected prostate cancer recurrence
(d) Detection of estrogen receptor (ER)-positive lesions in breast cancer

93. Which statement about ^{177}Lu-lutetium DOTATATE (Lutathera®) is false?

    (a) It is a radiolabeled somatostatin analog.
    (b) It is used to treat GEP-NETs (gastroenteropancreatic neuroendocrine tumors).
    (c) The beta emission from ^{177}Lu induces cellular damage.
    (d) The recommended dosage is weight based.

94. ^{131}I-iobenguane (Azedra®) is used for the treatment of?

    (a) Metastatic pheochromocytoma or paraganglioma
    (b) Thyroid cancer
    (c) Prostate cancer
    (d) Leukemia

95. Which of the following is NOT used to estimate β-amyloid neuritic plaque density?

    (a) Florbetaben
    (b) Florbetapir
    (c) Fulvestrant
    (d) Flutemetamol

96. In PET/CT imaging, which of the following corrections makes the biggest difference in image values?

    (a) Scatter correction
    (b) Attenuation correction
    (c) Randoms
    (d) Dead time

Appendix 2: Answers to Chapter 2

20

1. (d) ^{201}Tl decays by electron capture with gamma emissions of 0.135 MeV and 0.167 MeV.
2. (c) Stannous chloride is a reducing agent which changes the valence state of Tc in pertechnetate. It has no effect on either the amount of Al^{+3} or the radiation dose.
3. (b) The half-life of ^{131}I is 8.06 days. Since the time elapsed is about two half-lives, the original activity would be halved twice (50 mCi/2 = 25 mCi, 25 mCi/2 = 12.5 mCi).
4. (c) Dose should be 32 kg × 0.2 mCi/kg = 6.4 mCi.
5. (d) To find the effective half-life, we use the formula

$$T_e = \frac{T_p \times T_b}{T_p + T_b}.$$

$$T_e = \frac{12 \times 6}{12 + 6} = \frac{72}{18} = 4\text{h}$$

6. (a) T_p is used to abbreviate physical half-life as in the formula in the answer to Question 5 in this chapter.
7. (b) The physical half-life is fixed for any radionuclide and is the time necessary for the activity to be reduced to half its current activity. The biologic half-life is the time it takes for the body to eliminate half of the compound administered.
8. (b) Activity at time t = original activity × decay factor.
   $A_t = A_0 \times \text{DF} = (310\text{ mCi}) \times (0.618) = 191.6.$

E. Mantel et al., *Nuclear Medicine Technology*,
https://doi.org/10.1007/978-3-031-26720-8_20

9. (c) The physical half-life for ^{99m}Tc is 6 h. Since one half-life has elapsed, the original activity must be multiplied by 0.5.
10. (c) First, it is necessary to multiply 4 mL by 50 mCi/mL to arrive at the total activity in the diphosphonate kit = 200 mCi. Then the total volume is calculated (16 mL solution + 4 mL ^{99m}Tc = 20 mL). The specific concentration of the kit is activity/volume: 10 mCi/mL (200 mCi/20 mL). Then one uses the formula below:

    Required volume = Activity desired/specific concentration

$$\frac{20\,\text{mCi}}{10\text{mCi}/\text{mL}} = 2 \quad \text{mL}$$

11. (b) Activity at time T = Original activity × Decay factor

    $A_t = A_0 \times \text{DF}$ (60 mCi × 0.944 = 56.6).
12. (b) Following the addition of pertechnetate, sulfur colloid is heated in a shielded, boiling water bath. During this time, the ^{99m}Tc is rapidly incorporated into the sulfur colloid particles. Albumin colloid, diphosphonate kits, and MAA do not need to be heated during preparation.
13. (b) Al^{+3} in ^{99m}Tc eluate is an example of chemical impurity. Al^{+3} ions from the alumina column of the molybdenum generator must be less than 10 μg Al^{+3}/mL of eluate, the limit set by the US Pharmacopeia. This chemical impurity may result in reduced image quality due to poor labeling.
14. (d) The Nuclear Regulatory Commission has set a limit of ^{99}Mo in ^{99m}Tc eluate at 0.15 μCi/mCi of ^{99m}Tc at the time the dose is administered. The US Pharmacopeia regulates this radionuclidic impurity as well.
15. (b) Radionuclidic impurity is the activity present in the form of an unwanted radionuclide. In the case of ^{99}Mo in ^{99m}Tc eluate, this results in increased radiation dose and reduced image quality. Radiochemical impurity is the presence of the radionuclide in chemical forms other than that desired, for instance, the presence of free pertechnetate in a prepared kit of ^{99m}Tc sulfur colloid. Chemical impurity refers to the presence of other, nonradioactive chemicals in the sample, i.e., Al^{+3} in ^{99m}Tc eluate.

16. (c) 10 μg Al^{+3}/mL of eluate is the limit set by the US Pharmacopeia.
17. (d) The upper portion of the strip is the solvent front, and the lower portion is the origin (where the radiopharmaceutical being tested is introduced), but without further information about the chromatography kit, one cannot say what they specifically indicate.
18. (a) Specific concentration is 140 mCi/23 mL or 6.1 mCi/mL. Required volume is equal to the activity desired divided by the specific concentration (5 mCi/6.1 mCi/mL = 0.8 mL).
19. (b) Specific concentration is activity/volume. Since one half-life has elapsed, the specific concentration in the vial is 70 mCi/23 mL = 3.04 mCi/mL assuming that nothing has been withdrawn from the vial in the meantime. Required volume is activity desired divided by specific concentration (5 mCi/3.04 mCi/mL = 1.6 mL).
20. (a) Original activity is multiplied by the decay factor to give the activity at 10:00 a.m. (40 mCi × 0.794 = 31.76 mCi). Activity divided by the volume gives the specific concentration (31.76 mCi/5 mL = 6.35 mCi/mL). If we assume that a 4 mCi dose is desired, we must divide the activity desired by the specific concentration to obtain the volume needed (40 mCi/6.35 mCi/mL = 0.63 mL).
21. (b) Particles 10 μm or larger will be trapped by the capillaries in the lung, which measure 7–10 μm. Particles are formed which measure 5–100 μm, but most are in the range of 10–30 μm. An adult should receive 100,000–500,000 particles, which will occlude less than 1 in 1000 of the capillaries, in general.
22. (d) To find the effective half-life, we use the formula:

$$T_e = \frac{T_p \times T_b}{T_p + T_b}$$

Since the biologic half-life is in the range of 2–4 h, the equation is solved twice, first using two and then four to obtain the effective half-life of 1.5–2.4 h.

23. (c) ^{99m}Tc sulfur colloid is formed when elemental sulfur condenses in a heated solution forming colloid particles that incorporate ^{99m}Tc in the +7 valence state.
24. (a) See explanation to Question 15 in this chapter.
25. (b) Effective half-life must be shorter than either physical half-life or biologic half-life. See equation to calculate effective half-life in the answer to Question 22 in this chapter.
26. (d) EDTA is a chelating agent that will sequester the Al^{+3} ion, thereby helping to prevent aggregates from forming.
27. (c) Although package inserts state that doses should not be used after 6–8 h following kit preparation (depending on the particular kit used), most imaging departments prefer to use them within 4 h to obtain the best image quality.
28. (b) The appropriate particle size of sulfur colloid is 0.3–1.0 μm, which allows them to be phagocytized by the Kupffer cells of the liver.
29. (c) ^{99m}Tc human serum albumin has the smallest particle of the compounds listed, making it useful for applications like lymphoscintigraphy.
30. (a) It is not necessary to heat ^{99m}Tc albumin colloid during preparation, it is more expensive, and the recommended dose is comparable for both radiopharmaceuticals.
31. (a) In the absence of a shunt, 95% of ^{99m}Tc MAA particles injected will be trapped in the lung capillaries. In the event of a shunt, however, particles enter the left ventricle and the arterial blood. This results in the activity from the ^{99m}Tc MAA in the brain and lungs.
32. (d) Since there is no precalibration factor available, we can use the decay factor and rearrange the formula $A_t = A_0 \times \mathrm{DF}$ (see explanation to Question 11 in this chapter) to read $A_0 = A_t/\mathrm{DF}$ and solve for the original activity (22 mCi/0.707 = 31.1 mCi).
33. (d) Precalibration factors will always be greater than 1.0, since the activity precalibrated will be greater than the activity at the time of radiopharmaceutical administration. Similarly, decay factors will always be less than 1.0, since the activity is decreasing. We can use decay factors to solve precalibration problems as in the solution to Question 32, above.

34. (e) Some radiopharmaceutical doses are calculated according to weight. Clark's formula allows for the calculation of a pediatric dose by comparing the weight of a pediatric patient to an average adult weight (pediatric dose = (patient's weight in lb. × adult dose)/150 lb). A patient's body surface area can be calculated, and the appropriate fraction of the adult dose can be either calculated or found on a body surface area table. Talbot's nomogram is a calculated table relating body weight to surface area and the appropriate percentage of the adult dose. If the child's weight is known, the percentage can be found on the table and the dose calculated.
35. (c) First, specific concentration is calculated (820 mCi/10 mL = 82 mCi/mL). Since 41 mCi will be added, 41 mCi is divided by 82 mCi/mL (volume required = desired activity/specific concentration) to find a volume of 0.5 mL. Since 4–10 mL is the recommended total volume, at least 3.5 mL of diluent must be added.
36. (c) It is necessary to multiply 20 mCi by 1.259 (precalibration factor for 2 h, in Table 2.1) to obtain the answer, 25.18 mCi.
37. (b) If the decay factor needed is not available, it can be calculated by multiplying the decay factors for times that add together to make the elapsed time. In this case, the decay factor for 7 h is not available, so the decay factors for 3 and 4 h are multiplied to find the decay factor (0.707 × 0.630 = 0.445).
38. (a) Precalibration factors for each of the doses must be used. At 6:00 a.m., the dose of 10 mCi for 8:00 a.m. requires 12.59 mCi (10 mCi × 2 h decay factor of 1.259 = 12.59), the dose for 9:00 a.m. requires 14.14 mCi (using decay factor 1.414), and the dose for 10:00 a.m. requires 15.87 (decay factor 1.587). The three precalibrated doses are added together to obtain a total of 42.6 mCi as the minimum activity to be added to the kit.
39. (b) The counts for technetium in the desired form are expressed as a percentage of the total counts (258,000/(258,000 + 55,000) × 100), and the result is 82%. The lower limits of radiochemical purity differ according to radiopharmaceutical but typically are 90% and above, so this cannot be used.

40. (d) See answer to Question 39 in this chapter.
41. (c) The counts in the undesired form are expressed as a percentage of total counts to find radiochemical impurity (55,000/(258,000 + 55,000) × 100 = 17.6%).
42. (b) Because of a shorter half-life (6.01 h vs. 66.7 h), the technetium has decayed to a greater extent than has the molybdenum, and the breakthrough now exceeds the maximum allowed by the NRC, which is 0.15 μCi ^{99}Mo per mCi ^{99m}Tc.
43. (c) First, the activity at 9:00 a.m. has to be determined by multiplying the original activity by the 2 h decay factor (32 mCi × 0.794 = 25.4 mCi). The specific concentration is 12.7 mCi/mL (25.4 mCi/2 mL), and the desired dose must be divided by the specific concentration to find the volume to be retained in the syringe (11 mCi/(12.7 mCi/mL) = 0.87 mL to be retained). Therefore, 1.13 mL must be discarded (2 mL – 0.87 mL).
44. (d) The formula for decay calculation using half-life can be used:

$$A_t = A_0 e^{-0.693(t/\text{half life})}$$

where A_t is the activity at time t, A_0 is the original activity, and t is the elapsed time:

$$A_t = 5\,\text{mCi} \times e^{-0.693(2/6)}$$

$$A_t = 5\,\text{mCi} \times e^{-0.231}$$

$$A_t = 5\,\text{mCi} \times 0.794$$

45. (b) In order to find the number of particles, first it is necessary to know the volume of the 4 mCi dose. The specific concentration is 10 mCi/mL (50 mCi divided by 5 mL), so the required volume is 4 mCi/10mCi/mL = 0.4 mL (activity desired divided by the specific concentration). If there are 950,000 particles in 1 mL, there are 380,000 in the dose (particle concentration × volume of dose = particles in dose). Since most patients should receive 100,000–600,000 particles, this amount is acceptable.

46. (a) After 1 h the number of particles will remain unchanged, but the activity will decrease, so the number of particles per mCi will increase.
47. (a) Various sources state different particle numbers for patients with pulmonary hypertension, ranging from 60,000 to 200,000. The Society of Nuclear Medicine Procedure Guidelines recommend 100,000–200,000 particles. It is optimal to give the minimum number of particles, so the best choice from the answers given is 0.25 mL, which will give contained particles of 150,000 (600,000 particles per mL × 0.25 mL = 150,000 particles).
48. (b) Bacteriostatic sodium chloride will increase the oxidation products and can affect the radiochemical purity and distribution of the tracer.
49. (c) If there is also activity in the thyroid, the suspicion that the gastric activity is the result of free pertechnetate is supported. Providing the radiologist with a thyroid image will thus aid the interpretation of the scan.
50. (a) 1 mCi is equal to 37 MBq, so to convert mCi to MBq, multiply mCi by 37. The Becquerel (Bq) is the SI unit (International System of Units) of activity and is defined as one disintegration per second.
51. (b) Liver uptake on a bone scan can be the result of suboptimal pharmaceutical preparation from excessive aluminum ions in the eluate as well as excessive stannous ions, which may form tagged tin colloids that will be sequestered in the liver. It may also be the result of hydrolyzed, reduced ^{99m}Tc in the prepared kit. Additionally, various liver neoplasms also take up ^{99m}Tc-labeled phosphates, or the patient may have had a recent liver spleen scan. Lung uptake, if diffuse, could be due to metastatic pleural effusion. Thyroid and gastric activities are usually the result of free pertechnetate.
52. (d) Radiochemical impurity is the activity that is present in forms other than the desired form and will affect the biodistribution of the radiopharmaceutical. Introduction of air or water into the preparation vial can result in radiochemical impurity. During kit preparation, following addition of the

isotope, an equivalent amount of gas may be withdrawn into the syringe to normalize the pressure in the vial.

53. (b) The process of injecting air into the reaction may result in spillage and contamination of work areas and doing so will reduce the stability of the prepared radiopharmaceutical. Swabbing the vial septum maintains sterility.
54. (a) The Sievert (Sv) is the SI unit (International System of Units) of equivalent dose, and 1 rem is equal to 0.01 Sv, or 10 mSv. Therefore, multiply 15 rem by 10 mSv to find 150 mSv.
55. (b) The addition of hetastarch facilitates leukocyte labeling by assisting the settling of erythrocytes.
56. (d) Since kits are usually prepared to contain multiple patient doses, the patient name or ID number is not noted on the vial. Date, time, lot number, concentration, and volume should be noted on the vial. NRC licensees must maintain records regarding the dispensation of the vial contents, and this is where patient name and ID, administering technologist, isotope and radiopharmaceutical, activity of both prescribed and prepared dosages, and date and time of administration are listed.
57. (a) Labeled platelets may localize in thrombosis and cause a false-positive interpretation. If leukocytes are damaged during labeling, their normal function may be disrupted, resulting in a false-negative interpretation. All cell types will be labeled, so leukocytes must be labeled with care.
58. (d) This is true for most radiopharmaceuticals labeled with ^{99m}Tc; one exception is sulfur colloid, which uses unreduced ^{99m}Tc. In the other ^{99m}Tc radiopharmaceuticals, most of the ^{99m}Tc should be bound. The other forms are radiochemical impurity.
59. (a) The total volume is 502.5 mL, and 10 μCi is divided by that volume to find the concentration of 0.0199 μCi/mL.
60. (c) In transient equilibrium, the parent isotope has a somewhat longer half-life than the daughter isotope (^{99}Mo: T½ = 66 h, ^{99m}Tc: T½ = 6 h). The combined activity rises initially, then transient equilibrium is reached, and there is slightly more activity present in the daughter. If, instead, the

parent has a much longer half-life than the daughter, the daughter isotope activity increases until the activity of the daughter equals that of the parent. This is called secular equilibrium. If the daughter isotope has a longer half-life than the parent, no equilibrium is reached.

61. (b) Because a radiopharmaceutical may not contain more than 0.15 μCi of ^{99}Mo per mCi of ^{99m}Tc, 630 mCi must be multiplied by 0.15 μCi/mCi = 94.5 μCi.

21 Appendix 3: Answers to Chapter 3

1. (b) The Department of Transportation is the controlling authority for the packaging and transport of all hazardous materials.
2. (d) All the use of investigational pharmaceuticals is regulated by the Food and Drug Administration.
3. (b) The first priority in the event of a spill is to contain the contamination, i.e., to keep it from being spread. In this case, that would be accomplished by removing the clothing and storing it until the activity has decayed to background level.
4. (c) As in Question 3, the contamination must be contained as a first priority. Using a chelating agent may change the chemical structure of the substance spilled but will not affect the radioactivity of it. The Radiation Safety Officer should be notified, but first access to the area of the spill must be restricted.
5. (a) By doubling the distance, the intensity is reduced to one-quarter of the original level, as shown in this formula: $(I_1)(d_1)^2 = (I_2)(d_2)^2$, where I_1 is the original intensity, I_2 is the new intensity, d_1 is the original distance from the source, and d_2 is the new distance from the source. Substituting $2d_1$ for d_2, the $(d_1)^2$ terms cancel out, resulting in $I_2 = ¼\, I_1$.
6. (c) The best and simplest way to decrease exposure to a visitor would be to increase the distance from the patient. If the exposure rate at 1 ft from the patient is 3 mrem/h, and we

E. Mantel et al., *Nuclear Medicine Technology*,
https://doi.org/10.1007/978-3-031-26720-8_21

move the visitor to 4 ft from the patient, the exposure rate at the new distance is given by the equation in the answer to Question 5 of this chapter: $(I_1)(d_1)^2 = (I_2)(d_2)^2$. Plugging in 3 mrem/h for I_1, 1 ft for d_1, and 4 ft for d_2 yields $I_2 = 0.19$ mrem/h.

7. (b) Beta emitters can be effectively shielded by a few millimeters of plastic or Lucite. If shielded with lead, bremsstrahlung radiation will be produced from the slowing beta particles.
8. (b) The NRC annual dose limit permitted to the lens of the eye is 15 rem. The skin of extremities is permitted to 50 rem, and the whole-body total effective dose equivalent is limited to 5 rem (10 CFR § 20.1201).
9. (c) Gloves should be used when administering any pharmaceutical. Had the question stated the *radiopharmaceutical*, then a lead syringe shield also would be needed unless the radiopharmaceutical was a beta emitter.
10. (b) Wipe tests are used to detect removable contamination from surfaces such as packages, floors, and counters and are achieved by wiping 300 cm^2 of the area in question with a dry wipe and then counting the wipe along with a background sample in a well counter.
11. (a) DOT labeling categories are as follows: Radioactive White I, limited to packages with a dose rate of 0.5 mrem/h or less at the surface; Radioactive Yellow II, limited to packages with a dose rate of greater than 0.5 mrem/h but less than 50 mrem/h at the surface; and Radioactive Yellow III, for packages with a dose rate of greater than 50 mrem/h but less than 200 mrem/h at the surface. Choice (d) is incorrect because it does not give a dose *rate*.
12. (c) The gray (Gy) is an international unit (SI) measuring absorbed radiation dose and is equal to 100 rad. The millicurie (mCi) and the Becquerel (Bq) measure radioactivity according to disintegrations per time. The Becquerel is equal to 1 disintegration per second (dps), and the mCi is equal to 37 MBq.
13. (a) This can be solved using the inverse square law (see explanation for Question 5) to find 25 mrem/h. Or one can recog-

nize that the distance has been doubled, so the intensity is reduced to one-fourth of the original.

14. (d) ALARA is the mnemonic for the NRC's radiation protection philosophy that one should keep radiation exposure *as low as reasonably achievable*.
15. (b) Time, distance, and shielding are the most important factors to consider when attempting to reduce exposure to radiation.
16. (d) Airborne radiation, such as aerosols or gases, should be administered in rooms that are at negative pressure to surrounding areas. Depending on the amount of activity and the length of time it is present, special posting may be required (10 CFR § 20.1902).
17. (d) See explanation for Question 5.
18. (d) A sign reading "Caution Radioactive Materials" is required wherever radioisotopes are used. The sign reading "Caution Radiation Area" sign is required if the dose rate exceeds 5 mrem/h at 30 cm from the source, so no change in signposting is needed. However, the spill should be contained, so restricting access to the area until the activity has decayed to background is the best course of action.
19. (a) All technologists should use lead pigs and syringe shields when working with radiopharmaceuticals that emit gamma rays.
20. (c) OSHA requires that employees are advised at least yearly about their exposure records.
21. (b) According to the NRC regulations for signposting, the sign should be changed to "Caution Radiation Area" whenever exposure level is >5 mrem/h at 30 cm but a simpler solution would be to store the source in a shielded container.
22. (a) Time, distance, and shielding are the most important factors in reducing exposure; in this case, the simplest solution is to increase the distance to the source of the radiation.
23. (a) A package with such a label must have an exposure rate that does not exceed 0.5 mrem/h at the package surface, and there must be no detectable radiation at 1 m from the package.
24. (c) The transport index is determined by multiplying the maximum radiation level in mSv per hour at 1 m from the external surface of the package by 100. Because 1 mSv = 100 mrem, the transport index is also the maximum radiation level in mrem per hour at 1 m.

25. (e) The sign "Caution Radioactive Area" is posted wherever radioactive materials are used or stored. Unless patients are being injected in the reception area, a sign is not needed.
26. (b) Internal exposure may result from absorption, inhalation, or ingestion of radioactive materials. Using tongs to transfer a vial will not affect the possibility of inhalation or ingestion, and it would only affect the possibility of absorbing radioactive material through the skin if the technologist was not gloved. Using tongs is, however, a good practice to reduce exposure to external radiation as it increases the distance to the source.
27. (d) Personnel exposure is measured by all of the choices except the Geiger-Müller counter, which is a gas-filled ionization chamber. Because of dead time, it is typically used to measure small amounts of radiation.
28. (c) Cartons that have been used to ship radioactive material must have radiation labels removed or defaced before disposal. If wipe tests show contamination, it must be regarded as radioactive waste and disposed of according to the relevant regulations.
29. (b) The half-value layer (HVL) is the thickness of a material required to reduce the radiation intensity to half. The 13-mm-thick vial contains five half-value layers, so the intensity will be reduced in half five times (from 100 to 50 to 25 to 12.5 to 6.25 to 3.1 mR/h).
30. (e) NRC licensees are required to prepare and maintain reports of area surveys for 3 years. A list of isotopes used in the area is not a required part of the report.
31. (b) See explanation for Question 21. The sign reading "Caution: High Radiation Area" is not required until the dose rate exceeds 100 mrem/h.
32. (b) See explanation for Question 5.
33. (e) A medical event or other events should be reported to the NRC by telephone before the end of the next calendar day and in writing within 15 days. The referring physician should be notified within 24 h. Records of the event should be maintained for 5 years.
34. (e) A medical event can involve the wrong patient, the wrong radiopharmaceutical, the wrong route of administration, or a

dose that differs by more than 20% of the prescribed dose or falls outside the prescribed dose range. The key requirement to be a medical event is that the dose administered differs from the prescribed dose or dose that would have resulted from the prescribed dosage by more than 0.05 Sv (5 rem) effective dose equivalent, 0.5 Sv (50 rem) to an organ or tissue, or 0.5 Sv (50 rem) shallow dose equivalent to the skin. See 10 CFR § 35.3045 for details.

35. (a) Medical event records must be kept for 5 years. Records of patient dosage, area survey records, and records of instructions given to lactating females must be kept for 3 years.
36. (b) The NRC states that licensees who use ^{99}Mo-^{99m}Tc generators must keep records of the concentration of molybdenum in eluate for 3 years (10 CFR § 35.2204).
37. (b) The total dose limit for a fetus is 0.5 rem (10 CFR § 20.1208).
38. (d) If the radioactive material has a half-life <120 days, the material can be disposed of as ordinary trash after decaying in storage. Before disposal, the material must be surveyed and not be greater than background radiation (10 CFR § 35.92). Records must be kept of this for 3 years (10 CFR § 35.2092) and must include the radionuclide, the date of disposal, the equipment used to survey the material, the background dose rate, and the dose rate at the surface of the container used for disposal, as well as the name of the person disposing the material.
39. (a) Because ^{90}Y is a beta emitter, Lucite is an effective shielding material.
40. (b) The dose is 0.4 mCi/kg and is not to exceed 32 mCi. In general, patients may be released from hospitalization based on measured dose rates or on the basis of activity administered or retained. The activity below which a patient may be released differs according to the isotope (e.g., 33 mCi for ^{131}I). However, for ^{90}Y, ^{32}P, and ^{89}Sr the NRC does not require hospitalization because the exposure to the public from doses normally used is minimal.

Appendix 4: Answers to Chapter 4

22

1. (d) COR stands for center of rotation.
2. (a) The COR calibration is run to ensure that the axis of rotation is in the center of the computer matrix for all projections, and a separate COR calibration should be run for every collimator. The COR calibration should be checked as often as the scanner manufacturer recommends, which is often at least weekly and preferably each day that tomography will be performed. While the computer can correct for misalignments, if the COR needs frequent adjustment service may be needed.
3. (c) Flood field nonuniformities can cause artifacts in images, and this is a bigger issue in SPECT than in planar imaging because image reconstruction amplifies nonuniformities. Therefore, while 3–5% uniformity is acceptable in a correction flood for planar imaging, SPECT requires 1% or less. Due to Poisson statistics an average of 10,000 counts per pixel is necessary to achieve an uncertainty of 1%. If the image matrix is 64 x 64 = 4,096 pixels, then you need to acquire ~41 million counts. If the matrix is 128 × 128, the number of counts required is four times greater.
4. (c) The converging collimator magnifies the area imaged; the diverging collimator minifies it.
5. (b) Sealed sources must be leak checked during an inventory conducted every 6 months. Leak checks are made by wipe tests of the sealed sources.
6. (c) According to the NRC, dose calibrators with accuracy or constancy errors >10% must be repaired or replaced.

E. Mantel et al., *Nuclear Medicine Technology*,
https://doi.org/10.1007/978-3-031-26720-8_22

7. (d) Accuracy testing measures the ability of the dose calibrator to assay two calibrated reference sources with different principal photon energies and should be performed upon installation and annually thereafter. The NRC's suggested trigger level is a 5% difference between the measurement and the reference source's known activity. If the difference exceeds 10%, the dose calibrator must be repaired or replaced. The NRC requires that accuracy records be kept for 3 years.
8. (b) Linearity tests the ability of the dose calibrator to indicate the correct activity over the range of use between the highest dose administered to patients and 10 μCi. This can be performed over a 24-h period, using measurements and decay calculations for the isotope, or can be performed quickly using lead shields with a known attenuation. A linearity test should be performed at installation and at least quarterly thereafter. The NRC requires that records be kept for 3 years.
9. (a) Constancy is the ability of the dose calibrator to reproduce measurements of a source over a long period of time and should be checked with a dedicated check source at the beginning of each day of use. The NRC requires that records be kept for 3 years.
10. (b) Because the collimator is removed for intrinsic uniformity checks only a small amount of ^{99m}Tc is needed. Flood field uniformity should be checked daily.
11. (a) A Cobalt-57 sheet source is more expensive than a ^{99m}Tc tank source but it does not require any preparation, and due to its long half-life (272 days) it can be used for several years. Insufficient mixing of a fluid-filled source will result in suboptimal flood field images that can mask actual uniformity issues, and there is also the possibility of contamination with this method.
12. (a) Geometry testing is performed by measuring the same amount of activity in varying volumes. This should be checked upon installation of the dose calibrator, and the records should be kept for 3 years.
13. (a) When the distance between the patient and collimator is decreased, the system resolution improves, not only for

parallel-hole collimators but also for diverging, converging, and pinhole collimators.

14. (c) With parallel-hole collimators, the image size does not vary with distance. Although the image will not change in size when the patient-to-collimator distance decreases, the spatial resolution will improve.
15. (c) When a gamma ray interacts with a scintillation crystal such as thallium-activated sodium iodide, thousands of electrons are promoted to higher orbits. As they relax and fall back to their original positions, their excess energy is released as light.
16. (a) Many emitted photons will be attenuated in the patient, deflected away from the detector, or absorbed in the collimator, so most photons emitted do not contribute to the final image.
17. (d) A pinhole collimator produces high-resolution images but suffers from low sensitivity because a very low percentage of the photons incident on the collimator will pass through the pinhole.
18. (a) Geiger–Müller (GM) survey meters are more sensitive than ionization chamber survey meters by a factor of ~10. This allows them to detect spills with very low levels of contamination.
19. (b) The PM tube may have slipped out of position when the detector faced upward and dropped back into place when the detector head was face down, causing a defect that was only observable in the first image.
20. (b) When photons interact with the crystal simultaneously, or nearly simultaneously, they may be perceived as a single event with the energies summed.
21. (c) Dead time is the time required after the absorption of a photon before the next photon can be detected.
22. (b) The amount of light produced in the crystal is small and is converted into an electrical signal which is multiplied in the photomultiplier tube.
23. (a) In a scintillation camera the pulse height analyzer determines the energy that each gamma ray has deposited in the crystal by measuring the summed output of all photomultiplier tubes (known as the Z signal). This information is used

to limit the range of photon energies accepted into the final image, so that fewer Compton-scattered photons will be included.

24. (b) Multiple PHAs are useful for imaging radionuclides with multiple peaks. ^{67}Ga has four photopeaks (92, 185, 300, 394 keV), ^{99m}Tc and ^{133}Xe each has one photopeak, and ^{111}In has two (172, 245 keV).
25. (b) A symmetric window will be divided equally above and below the photopeak. Fifteen percent of 140 keV means that 21 keV around the photopeak will be accepted, 10.5 keV above and 10.5 keV below. The 158 keV photon is 18 keV above the photopeak and will be rejected.
26. (d) Compton scatter occurs when a photon interacts with a loosely bound outer-shell electron. The photon transfers some of its energy to the electron and is deflected in a new direction with lower energy. The more electrons that a photon encounters, the more likely it is that this process will occur. Therefore, photons traveling through air rarely experience Compton scattering, while photons traveling through patients do.
27. (b) Spatial resolution refers to the ability of a system to discriminate between two structures located close to each other and is typically measured in millimeters. To assess spatial resolution phantoms containing either lead bars that vary in thickness and distance from one another or multiple holes in lead sheets are imaged.
28. (a) Energy resolution, like intrinsic spatial resolution, is primarily dependent on statistical fluctuations in the distribution of photons among photomultiplier tubes from one scintillation event to the next. Higher energy gamma rays deposit more energy in the scintillation crystal, producing more secondary photons. With more photons, there is lower statistical fluctuation in their distribution (per Poisson statistics), so energy resolution is improved.
29. (d) The orthogonal hole phantom contains a lead sheet with holes in it and requires only one image to assess resolution. The other phantoms contain lead bars and spaces and require multiple images to assess resolution across the entire detector.

30. (d) The maximum in counts is 11,000, so full width at half maximum is the width at 5500 counts or 78 keV (701–623 keV), and energy resolution is solved by using the formula: % energy resolution = (FWHM/radionuclide photopeak) × 100. Applying the formula, energy resolution is found to be [(701 keV–623 keV)/662 keV] × 100 = 11.8%.
31. (b) The orthogonal hole phantom, although it does not contain lead bars, can still be used to assess linearity because holes are arranged in parallel lines.
32. (a) Sensitivity is expressed as counts per time per μCi. The background count rate must be subtracted from the measured count rate before dividing by the number of μCi: (53,125–405) cpm/150 μCi = 351 cpm/μCi.
33. (c) If removing the collimator and obtaining an intrinsic flood results in a significantly lower uniformity %, the problem could be either a damaged collimator or a poorly mixed tank source.
34. (c) See Questions 25 and 26 and their explanations. An asymmetric window can be used to exclude photons that have undergone Compton scatter.
35. (b) A 20% symmetric window around 364 keV means that the window extends from 364 – (10% of 364) keV below the photopeak to 364 + (10% of 364)) above the photopeak.
36. (c) Images obtained are called projections; these can be processed to create reconstructions. The detector head rotates in an arc, and the motion may be interrupted by azimuth stops during which imaging is performed.
37. (b) The distance between the sources is divided by the number of pixels between the activity peaks to find 0.681 cm (6.8 mm).
38. (d) Increasing the number of projections will reduce the star artifact. This problem was largely solved by systems that use filtered as opposed to simple back projection.
39. (a) In step-and-shoot mode, data acquisition stops while the camera head(s) move to the next azimuthal angle. In continuous mode, the camera(s) rotate continuously while data is acquired continuously, so data acquisition takes less time. However, the continuously acquired data must be interpolated

back into azimuthal steps, and that results in slightly worse spatial resolution. Or, to phrase it another way, step-and-shoot mode has better spatial resolution because there is no blurring caused by camera rotation during acquisition.

40. (a) Increasing the matrix size divides the same physical space into smaller "chunks" or picture elements (pixels). Therefore, spatial resolution should increase. However, because a 128 × 128 matrix has 4× more pixels than a 64 × 64 matrix, for the same imaging time each of its pixels will contain only 1/4 of the counts, resulting in noisier images due to Poisson statistics. Increasing the imaging time helps counteract this effect. The object that is emitting photons (the patient) is unchanged, so the count rate will not change.
41. (a) The pixel size for x is 6.73 mm/pixel (35 cm/52 pixels), and the size in the y direction is 6.36 mm/pixel, so they are within 0.5 mm of one another. This is critical for volume calculations to be considered accurate.
42. (c) This is a partial test of accuracy. Accuracy is the ability of the system to measure samples with different energies, so at least two different isotopes must be measured.
43. (a) Quality control records for dose calibrators should be kept for 3 years.
44. (b) A positron with a long range can travel millimeters away from the site of the radioactive decay that produced it, while a PET camera only detects where the positron finally slows down enough to interact with an electron and annihilate, not where it originated. Because images from radioisotopes with long positron ranges, like ^{82}Rb, have larger mismatches between where the positron annihilation is detected and the actual origin point of the positron, they have worse spatial resolution than images from radioisotopes with short positron ranges, like ^{18}F.
45. (a) Before using a GM meter each day the user should perform (1) battery test—check that the needle goes up into region marked BAT TEST and (2) constancy test—hold the detector up against a check source and compare the measurement (minus background) to the expected value.

46. (b) Geiger–Müller meters are designed to detect small amounts of contamination. Ionization chambers are not sensitive enough to use for contamination monitoring.
47. (b) Because of the low voltage there is no avalanche effect, and therefore no dead time.
48. (b) Thermoluminescent crystals are used for monitoring radiation exposure to occupational personnel. The Geiger–Müller counter uses an applied voltage from anode to cathode. Silver halide coats the radiographic film that was often used in nuclear medicine departments.
49. (c) Well counters best measure samples containing small amounts of radioactivity and are useful for counting in vitro studies, wipe tests, etc. Samples should have identical geometry, i.e., the volumes and containers used should be identical for all samples.
50. (b) For each time point multiply the original activity by the decay factor to calculate the expected value. If all measured values are within 10% of their expected value the dose calibrator can be put into service.
51. (a) Multiplying the original activity by the decay factor one finds that expected activity is 126 μCi; 10% of the expected activity must be added to and subtracted from the expected activity to find the range.
52. (b) The true activity (measured in the smallest volume) is divided by the measured activity to find the correction factor.
53. (c) Energy resolution is the ability of the system to reliably convert light to an electrical impulse, and it should be assessed annually. It requires a graph of the counts vs. the energy spectrum and knowledge of the full width of the curve at half maximum. It can be calculated using the following formula: % energy resolution = (FWHM/radionuclide photopeak) × 100. Applying the formula, energy resolution is found to be [(701 keV–623 keV)/662 keV] × 100 = 11.8%.

Appendix 5: Answers to Chapter 5

23

1. (b) RAM refers to random access memory, which is used for temporary data storage and is sometimes referred to as cache memory. ROM refers to read-only memory, and data stored there cannot be modified as easily as can data stored in RAM. A rad is the traditional unit for absorbed dose (an acronym for radiation absorbed dose), and REM is a sleep stage in which there is rapid eye movement.
2. (c) Magnetic tape is a low-cost option for data storage, but the data retrieval time is significantly longer than optical disk, floppy disk, or hard drive storage.
3. (a) A high-pass filter allows high frequencies to pass through and reduces lower frequency data from the resulting image. This can produce a noisier image with better edge definition.
4. (c) In addition to display on conventional film, nuclear medicine images can be displayed on CRTs and video monitors. Magnetic tape is a storage device rather than a display output.
5. (b) The array processor is used to perform filtering at high speeds. A buffer is a temporary storage area. ROM is read-only memory, and ADC is the abbreviation for analog-to-digital converter.
6. (d) In frame-mode acquisition, the x–y location of each event is determined and the corresponding pixel value in the image matrix is incremented by one. Acquisition is stopped after a preset number of events is recorded or the amout of time has

E. Mantel et al., *Nuclear Medicine Technology*,
https://doi.org/10.1007/978-3-031-26720-8_23

elapsed, and the pixel values are then stored. Data acquisition in frame mode requires less memory and has a higher acquisition rate than list mode, but the data may not be divided into different images at a later time. By contrast, in a list-mode acquisition the x–y location for each individual event is stored along with periodic clock signals. These time markers permit the events to be retrospectively reframed into shorter or longer duration images.

7. (a) List mode is more useful for gated first-pass studies because the data can be manipulated in many different ways to visualize anatomy and to generate time-activity curves. The disadvantage is that it requires more memory and has a lower acquisition rate than does frame-mode acquisition.
8. (b) DICOM stands for Digital Imaging and Communications in Medicine and is the international standard to transmit, store, retrieve, print, process, and display medical imaging information.
9. (a) For MR, CT, and PT images the transverse orientation is the standard view.
10. (c) The DICOM header contains details about the scanner and the images that it produced, along with patient identifiers such as patient name, MRN, and DOB. Because the DICOM header contains PHI, DICOM images must be stored and transmitted securely.
11. (d) In the Nuclear Medicine DICOM Standard, a set of images can be transferred as a multi-slice or multi-frame image where all slices and frames are contained in a single file.
12. (c) A Picture Archiving and Communication System (PACS) is a computer system used to store, distribute, and display digital images and reports. PACS also provides tools for advanced image manipulation and analysis.
13. (a) A Radiology Information System is a sophisticated database system customized for use in radiology. It typically facilitates patient management tasks such as patient registration and scheduling, and ordering exams. Once an exam is scheduled, the RIS can automatically prefetch the patient's relevant prior images from the PACS. The RIS then supplies the imaging modality with information about the patient such

as name, ID, and DOB, thus preventing data transcription errors. When the image data acquisition is complete the RIS is notified and the images are stored in the PACS. A RIS can also store financial records, process electronic payments, and automate billing.

14. (d) Filters may be applied before (preprocessing), during, or after (postprocessing) reconstruction. The filter that is most often used during reconstruction with filtered back projection is a modified ramp filter, commonly the Butterworth filter.
15. (b) Filters are useful for smoothing images, especially when there are not many counts. Spatial filters are applied to static images, while temporal filters are applied to dynamic images. A low-pass filter reduces high spatial frequencies, and a band-pass filter removes both high and low spatial frequencies.
16. (b) Single-emulsion film is used because only one side of the film is exposed to light.
17. (c) When there is a suspicious area on an image, whether hot or cold, the technologist is responsible for helping the radiologist determine the cause of the spot. This could involve cleansing the patient's skin and/or removal of clothing followed by reimaging, as well as taking images at additional angles (oblique or lateral views). If radioactive contamination is suspected, the technologist should survey the table and cushions once the patient departs.
18. (e) Labeling left and right sides and injection site will enable the radiologist to accurately interpret the study. Young patients will have normally increased activity in the epiphyses, so age should be provided. Old fractures and areas of prior surgery may cause hot spots that can be mistaken for new pathology, so this information should be provided as well.
19. (d) Film should be stored on its side, in the foil wrapper to reduce exposure to light and moisture, and at room temperature.
20. (d) Static on film should be avoided as it may cause artifacts and can come from any of the situations listed.
21. (d) If the developer is diluted, or not replenished as required, or if the temperature of the developer is incorrect films may

appear too light. All of the situations listed should be monitored.

22. (c) To develop film, the silver halide crystals in the emulsion that have been exposed to light during imaging are reduced to metallic silver. During fixing, those silver halide crystals that were not exposed to light are removed, and the silver metal is left behind. The developer is neutralized to stop the reducing activity.
23. (a) Filtered back projection is an algorithm used to reconstruct tomographic images following a SPECT study. Transverse images are reconstructed first; these are then used to generate sagittal and coronal images. Unfiltered back projection often results in star artifacts.
24. (a) A gray scale assigns the level of brightness to a pixel value, but does not affect the fidelity with which the image represents the original object. Spatial resolution is the detail or sharpness of an image. Contrast is the difference in intensity in parts of the image corresponding to different concentrations of radioactivity in the patient. Noise can be random or structured. Random or statistical noise causes the speckled appearance of NM images and is due to random statistical variations in the counting rate. Structured noise is nonrandom variations in counting rate and can arise from imaging system artifacts.
25. (b) ROIs can be any geometric shape (e.g., rectangle, circle) or can be drawn manually in an arbitrary shape.
26. (d) Contrast resolution is the ability to distinguish between differences in intensity in an image and is difficult to define because it depends on the human observer as much as the quality of the actual image.

Appendix 6: Answers to Chapter 6

24

1. (b) Ewing's sarcoma is a type of cancer most often found in children and young adults and is more common in males than females. Paget's disease is a chronic bone disease involving enlarged and deformed bones. Osteomyelitis is an infection of the bone or bone marrow. Osteoid osteoma is a benign bone tumor that is common in the appendicular skeleton. Osteoid osteoma demonstrates tracer uptake on blood pool images during a bone scan as opposed to most benign osseous neoplasms.
2. (c) Cellulitis often demonstrates as diffuse increased activity on early images, and activity decreases on delayed images. Osteomyelitis shows increased uptake in bone on early images, which continues to increase on delayed images.
3. (b) Free pertechnetate may be due to poor tagging of phosphates, or too much time may have elapsed between radiopharmaceutical preparation and injection.
4. (d) Stannous chloride reduces the valence state of technetium, thereby improving the tag efficiency.
5. (c) Planar bone scans can be performed using 10–30 mCi of ^{99m}Tc phosphates and phosphonates; higher doses may be indicated in obese patients and calculated based on weight (300–350 μCi/kg).
6. (b) Since most of the injected dose will leave the body via the urinary tract, the skin contaminated by urine is a common

E. Mantel et al., *Nuclear Medicine Technology*,
https://doi.org/10.1007/978-3-031-26720-8_24

problem. Metal on clothing will cause an attenuation artifact. If there is even a small amount of extravasation of the dose during injection, there will be a hot artifact at the injection site. A colostomy bag is the least likely to cause an artifact on bone scans.

7. (b) Bone mineral density, which is decreased in osteoporosis, can be evaluated using quantitative CT (QCT), dual photon absorptiometry (DPA), and dual X-ray absorptiometry (DXA). Detection and follow-up of metastases and differentiation of osteomyelitis and cellulitis are common indications for bone scanning. It is also used for evaluating avascular necrosis.
8. (d) Free pertechnetate demonstrated as gastric, salivary, and thyroid activity on bone scans may be the result of administering a dose from a kit that has been prepared more than 4 h prior to injection. Additionally, if air has been introduced into the reaction vial during labeling, there may be a poor technetium–phosphate tag.
9. (d) Hydration and frequent voiding after administration of a diphosphonate radiopharmaceutical reduce the radiation dose to the bladder since the radiopharmaceuticals are excreted through the urinary bladder. Patients should void immediately before imaging so that bladder activity does not obscure visualization of the pelvis.
10. (b) Malignant ascites may be seen as diffusely increased activity on bone scans. Malignant pleural effusion may be similarly seen in the chest.
11. (a) The fourth phase of imaging is usually completed approximately 24 h following injection; therefore, choices (b) and (c) are incorrect. This protocol may be necessary in patients with renal failure because soft tissue activity may persist.
12. (e) Focal hot spots on bone scans that appear to be in soft tissue, particularly on the medial aspect of the thighs, are often the result of radioactive contamination on clothing or skin from urine. Patients should remove clothing and wash the skin to remove the possible contamination.
13. (b) Decreased renal activity, together with diffusely increased bone uptake, especially in the axial skeleton, is often referred

to as a *superscan*. This is most often seen where there is diffuse metastatic involvement in the skeleton but may also be due to certain metabolic conditions. The *flare phenomenon* refers to the apparent worsening of bone metastases in the months following chemotherapy, which is actually due to increased uptake resulting from healing. Patients in renal failure often show overall increased soft tissue uptake.

14. (d) The *glove phenomenon* is markedly increased activity in the distal extremity and is due to an arterial injection.
15. (d) Prostate, lung, and breast cancers often metastasize to the bone, most commonly in the bones of the thorax, spine, and pelvis.
16. (a) ^{99m}Tc, following elution from an alumina column generator, exists primarily in the valence state +7. Stannous ions reduce the valence state, usually to +4, and this improves the tag efficiency.
17. (b) The appendicular skeleton refers to the appendages (which includes the femurs, phalanges, and radius; choices (a), (c), and (d)) and the bones of the pelvic and shoulder girdles. The skull, spine, most bones of the thorax, and some pelvic bones comprise the axial skeleton.
18. (f) See explanation to Question 17.
19. (c) It is believed that ions in the tracer exchange with those in the bone crystal, hydroxyapatite. Uptake appears to be related to osteogenic activity and skeletal blood perfusion.
20. (c) Pediatric bone scans show intense uptake in the area of the epiphyses, because the cartilaginous epiphyseal plates are the sites of longitudinal bone growth. When bone growth stops, the cartilage is replaced by bone.
21. (a) A three-phase bone scan is begun by administration of a bolus injection followed by imaging every 1–3 s, for at least 1 min. Choice (d) is incorrect because the dynamic images are not performed for a sufficient length of time.
22. (d) Bone marrow scans are commonly performed using ^{99m}Tc sulfur or albumin colloid because the marrow contains reticuloendothelial cells, which localize the colloid by phagocytosis.

23. (c) SPECT is useful for evaluating avascular necrosis. Pediatric cases may especially benefit from pinhole collimation because the epiphysis is small and there is nearby activity from the bladder and the acetabulum.
24. (b) Patients with sickle cell anemia often have splenic uptake on bone scans due to splenic infarcts. The radiopharmaceutical localizes in the spleen following infarct due to revascularization; if infarcts are repetitive, there may be mineral deposits which also cause uptake.
25. (e) Bone's osseous tissue contains calcium phosphate in the form of hydroxyapatite. Collagen provides some elasticity.
26. (d) Osteoblasts are cells that create collagen and strengthen bone; therefore, osteoblastic activity refers to formation of new bone.
27. (b) Osteoclasts are cells that destroy mineral tissue as part of a process called bone turnover. Therefore, osteoclastic activity refers to destruction and reabsorption of bone.
28. (d) The skeleton provides support and protection for organs, and the bone marrow is the site of production of leukocytes and erythrocytes.
29. (b) A large number of bone tumors develop in the first decades of life, for example, Ewing's tumor, osteosarcoma, and chondroblastoma. Many also occur in young adulthood, e.g., giant-cell tumors and primary bone lymphoma.
30. (c) Since radiopharmaceuticals used for bone scanning are excreted in urine, the bladder receives the highest radiation dose. Hydration and frequent voiding reduce the radiation dose to the bladder.
31. (b) When performing a three-phase bone scan of the upper extremities, the technologist should obtain venous access and then release the tourniquet and delay injection slightly to avoid increased uptake on the injected side.
32. (b) The technologist can position the patient directly adjacent to the detector head during spot planar imaging.
33. (c) It is important to check the patient's history and the referring physician's suspicions before injection, to ensure that areas of potential interest are not obscured by extravasation of the radiopharmaceutical.

34. (d) Being NPO, taking a cleansing enema, or refraining from taking thyroid medication will have no effect on bone scan images.
35. (a) The most common sites for bony metastases are the spine, thorax, and pelvis.
36. (e) Bone lesions are undetectable on radiographs until 30% or more of the bone's calcium content is lost; lesions can be seen on bone scans even when there is a slight loss of calcium. When there is suspicion of multifocal trauma, a bone scan can more efficiently survey the whole body than can conventional radiography.
37. (c) Radiolabeled sulfur or albumin colloid particles are phagocytized by the reticuloendothelial cells in the bone marrow.
38. (b) There are many reasons for soft tissue activity on bone scans. Of the choices listed, renal insufficiency is the most likely cause, although there can be diffuse soft tissue in the abdomen or chest in cases of metastatic ascites and metastatic pleural effusion.
39. (b) ^{99m}Tc HDP is used for bone scintigraphy and for evaluation of cardiac amyloidosis. ^{99m}Tc MDP is most commonly used for bone scan but not for cardiac imaging. ^{99m}Tc pyrophosphate (PYP) is most commonly used for evaluation of cardiac amyloidosis. Although a good bone seeking agent, ^{99m}Tc PYP is no longer used for bone scintigraphy. ^{201}Tl chloride is used for evaluation of myocardial perfusion.
40. (a) Rib fractures are often found in consecutive ribs. Bone metastasis is more likely to appear as linear uptake along the rib.
41. (d) The cold spot in the left proximal humerus, which is bordered by increased uptake, is most likely the result of surgically implanted metal. The cold area in the distal humerus is the result of shielding.

25 Appendix 7: Answers to Chapter 7

1. (b) The superior sagittal sinus is a dural venous sinus located in the midline. Tracer usually appears thereafter about 15 s signifying the venous phase of the blood flow study. In some cases, low-level activity in the sagittal sinus from the scalp can be mistaken for venous flow.
2. (b) ^{99m}Tc HMPAO (exametazime) enters the brain via the cerebral blood flow. It then crosses the blood–brain barrier and is metabolized to a form that cannot diffuse out of the brain. As a result, the Tc-99 m HMPAO activity in the brain represents blood flow and brain perfusion.
3. (b) Cerebrospinal fluid, which is produced in the choroid plexus and serves to cushion the brain and spinal cord, is about 99% water. It also contains a small percentage of plasma proteins.
4. (a) Cerebrospinal fluid acts as a shock absorber for the brain and spinal cord. It occupies the subarachnoid space and the cord's central canal.
5. (b) Images are ideally obtained anteriorly, with the patient upright whenever feasible. The neck should be included, and the top of the head should not be outside the FOV. However, in many cases, the condition of the patient makes this positioning impossible.

E. Mantel et al., *Nuclear Medicine Technology*,
https://doi.org/10.1007/978-3-031-26720-8_25

6. (c) HMPAO is hexamethyl propylene amine oxime and is also known as exametazime. ECD means ethylene l-cysteinate dimer and is also known as bicisate. MAG3 is mertiatide, and DTPA is diethylenetriaminepentaacetic acid.
7. (d) See explanation for Question 6.
8. (c) ^{111}In DTPA (or other radiopharmaceuticals) is injected by lumbar puncture into the subarachnoid space, the intrathecal space surrounding the spinal cord.
9. (c) The Society of Nuclear Medicine recommends 15–20 mCi of ^{99m}Tc DTPA for brain death scintigraphy. 10–30 mCi is recommended for ^{99m}Tc HMPAO and ^{99m}Tc ECD.
10. (a) Since various stimuli and cognitive functions will affect regional blood flow, the venous access should be achieved a few minutes before injection. The patient should then wait in an atmosphere of minimal stimuli (light, noise, pain, etc.).
11. (d) 15–30 mCi of ^{99m}Tc HMPAO or ^{99m}Tc ECD is injected for SPECT brain imaging.
12. (d) Ictal examinations are performed by injecting the radiopharmaceutical during or within 30 s following a seizure. Interictal examinations are those performed between seizures.
13. (a) Because HMPAO and ECD do not redistribute in the brain for at least an hour following intravenous injection, they are particularly useful for ictal studies, allowing an image of blood flow during seizure activity.
14. (d) HMPAO should be used within 4 h after kit preparation. ECD can be injected up to 6 h after labeling and clears more quickly from the blood. This rapid clearance results in increased target to background.
15. (b) Gray matter has much greater blood flow than white matter. For this reason, gray matter will have greater activity than white matter when performing these examinations with HMPAO or ECD.
16. (a) A dynamic study is usually obtained immediately following injection, and further imaging may be carried out at any time up to 2 h after injection. If brain-specific agents are used, images may be obtained after a delay of at least 20 min.

17. (c) When a radionuclide angiogram is used to confirm brain death, it may be difficult to differentiate scalp perfusion from intercerebral blood flow. An elastic band situated above the orbits can reduce the flow to scalp vessels.
18. (c) Metabolic refers to the building up of substances (anabolism) and the breaking down of substances (catabolism). FDG is a glucose analog and therefore is metabolized initially in the same way that glucose is. Unlike glucose, it is not further metabolized after localizing in the brain substance (because it is phosphorylated) and is therefore a useful imaging agent.
19. (a) In areas of normal perfusion, blood flow increases 3–4 times following the administration of Diamox, or acetazolamide, which causes vasodilatation. In patients with decreased regional perfusion, the Diamox challenge study may show a relative perfusion decrease compared with surrounding normal tissue. About half of the patients will experience side effects, and so all patients should be carefully monitored, although symptoms (light-headedness, facial flushing and numbness, and numbness in the fingers) usually resolve themselves in about 15 min.
20. (d) In a shunt patency study, the radiopharmaceutical is administered directly into the shunt reservoir or the tubing. Activity is then followed to diagnose a malfunction of the shunt. It is necessary to understand the particular type of shunt and its path to diagnose obstruction.
21. (e) Communicating hydrocephalus results when CSF resorption is faulty. Noncommunicating hydrocephalus is caused by obstruction of CSF flow. Shunts may be used to treat communicating hydrocephalus. CSF leaks may result from hydrocephalus (among many other things), but shunts are not used to treat the leak.
22. (d) ^{99m}Tc HMPAO, or exametazime, is a lipophilic radiopharmaceutical. It crosses the blood–brain barrier and then is metabolized to another form which does not diffuse out of the brain. The other choices only cross a compromised blood–brain barrier.
23. (a) ^{99m}Tc DTPA does not cross the blood–brain barrier and is useful for radionuclide cerebral angiograms. ^{99m}Tc HMPAO

and ^{99m}Tc ECD are nondiffusible tracer with persistent activity within the brain parenchyma, and thus can be used for SPECT studies of brain perfusion.

24. (a) When performing a radionuclide brain death scan, placing an elastic band around a patient's head just above the orbits during a radionuclide angiogram can reduce blood flow to scalp vessels, and thus help differentiate from cerebral vascular activity for a nondiffusible tracer such as ^{99m}Tc DTPA. This is not necessary for ^{99m}Tc HMPAO or ^{99m}Tc ECD. Both HMPAO and ECD cross the blood–brain barrier, are consequently metabolized within the brain parenchyma, and cannot diffuse out of the brain.
25. (d) Radionuclide angiography is commonly used to confirm brain death and may be performed using ^{99m}Tc pertechnetate, ^{99m}Tc DTPA, ^{99m}Tc GH, or the lipophilic radiopharmaceuticals ^{99m}Tc HMPAO or ^{99m}Tc ECD. ^{99m}Tc HMPAO or ^{99m}Tc ECD is preferred over others, as they provide brain perfusion information and SPECT images can be obtained.
26. (c) Activity in the Sylvian fissure and the basal cisterns is usually seen in early images (2–4 h) and cerebral convexities later (by 24 h). Visualization of the lateral ventricles is usually considered abnormal and may indicate hydrocephalus.
27. (c) Most labs use 0.5–1.0 mCi of ^{111}In DTPA.
28. (c) Pledgets are placed prior to radiopharmaceutical administration, removed 4–24 h later, and counted in a well counter. Blood serum is also sampled concurrently, and pledget/serum count ratios are calculated; ratios greater than 1.3–1.5 are generally considered abnormal. This information is evaluated together with images of the leak site.
29. (c) ^{99m}Tc DTPA is not useful for evaluating CSF dynamics in adults because of its shorter half-life. Children have faster flow, so ^{99m}Tc DTPA can be used in pediatric examinations.
30. (d) Minimizing patient-to-detector distance and immobilizing the patient are important considerations for all imaging examinations in nuclear medicine. Various stimuli can affect cerebral blood flow. The Society of Nuclear Medicine Procedure Guidelines recommend preparing the patient in a quiet room with low lighting, obtaining venous access 10 min

before injection, instructing the patient not to read or talk during the preparation period, and not interacting with the patient before, during, and immediately following injection to avoid increased uptake in relevant areas of the cortex.

31. (c) Thyroid blockage by SSKI or Lugol's solution should be done prior to the study. No NPO or coffee/tea abstinence is required. Voiding does not help as the imaging is limited to the head.
32. (c) DaTscan® has high binding affinity for presynaptic dopamine transporter, which is decreased in patients with Parkinsonian syndromes.
33. (d) It is important to keep the head in proper position and close to the camera.
34. (d) There is no need for NPO and no limit for blood sugar for Amyvid® PET study, and arms should be down for head imaging.
35. (c) Amyloid PET detects brain beta-amyloid deposition, a hallmark of Alzheimer's disease.
36. (e) Coffee or tea may stimulate the brain activity. Brain FDG-PET imaging can be obtained as soon as 30 min after tracer injection.
37. (b) Amyloid PET imaging is most helpful in patients with clinically suspected Alzheimer's disease, but not needed in patients with clinically diagnosed Alzheimer's disease.
38. (d) ^{11}C-Pittsburgh Compound B (PIB) is the initial investigational candidate for imaging Alzheimer's disease.
39. (d) No special patient preparations regarding diet and medicine for amyloid PET study.
40. (c) ^{18}F-florbetapir, dose of 370 MBq (10 mCi), waiting period of 120 min.
41. (b) Amyloid PET imaging is most helpful in patients with clinically suspected Alzheimer's disease. However, positive findings on amyloid PET cannot be used to confirm the diagnosis or to determine the severity of Alzheimer's disease.
42. (a) FDG-PET is used for evaluation of memory disorder. Different from amyloid PET imaging, FDG-PET imaging detects abnormal brain metabolism to help differentiate different types of dementia.

Appendix 8: Answers to Chapter 8

26

1. (a) The QRS complex represents depolarization of the ventricles. The P wave represents depolarization of the atria. The ST segment is representative of a pause between depolarization and repolarization of the ventricles. The T wave represents repolarization of the ventricles.
2. (d) Beta-blocker is not recommended for exercise stress test but is not an absolute contraindication.
3. (b) Following injection, the patient ideally continues to exercise for 30 s to 1 min to allow localization of the radiopharmaceutical in the myocardium, and imaging begins as soon as possible. Most sources state 10–15 min as the maximum time allowed before imaging is begun. Although 5 min would be better, it is not usually possible to remove the patient from the exercise situation and into imaging position within this time.
4. (d) For thallium, imaging should start at soon as possible, ideally within the first 5–10 min of exercise. The quality of repeat imaging is likely compromised due to tracer redistribution in the myocardium. As a result, the benefit of repeating image acquisition is questionable.
5. (c) The 45° LAO is, from the choices listed, the best choice for imaging when calculating the left ventricular ejection fraction.
6. (b) Ejection fraction is the portion of blood in the ventricle that is pumped out with a single heartbeat. Ejection fraction is

E. Mantel et al., *Nuclear Medicine Technology*, https://doi.org/10.1007/978-3-031-26720-8_26

calculated as a percentage of the counts at end diastole using the formula:

$$\frac{\text{End diastolic counts} - \text{End systolic counts} \times 100}{\text{End diastolic counts} - \text{Background counts}}$$

Because the counts given in the question are net counts, we do not need to subtract background. Using the formula given, 39% is the ejection fraction.

7. (b) Ejection fraction of the left ventricle is usually considered normal if above 50%, though some sources state 55%. Normal EF for the right ventricle is >40%.
8. (b) Without calculating, it can be seen that there is a greater difference in counts at the end of end diastole vs. end systole (counts at end systole are nearly a third of those at end diastole) in the lower curve than on the upper curve (where end systolic counts are roughly half of end diastolic counts). The ejection fraction calculated for the upper curve is 66%, and the ejection fraction represented by the lower curve is 48%.
9. (b) The radionuclide ventriculography can be used to assess wall motion but is not ideal for measuring wall thickness because camera resolution and partial volume effects limit the reliability of such measurements. Dyskinetic segments resulting from left ventricular aneurysms can be assessed by creating a paradox image (subtraction of the diastolic frame from the systolic frame), which may show increased activity in this case. Ventricular aneurysm may also cause parts of the wall to contract out of phase with the rest of the ventricle.
10. (b) The vertical long-axis projections have been created during a SPECT ^{201}Tl study. These images logically show the left ventricle from the septum to the lateral wall of the ventricle.
11. (b) If there are 65 bpm, each cardiac cycle lasts 0.92 s (60/65–0.92). If that cycle is divided into 24 images, each image represents 38 ms (0.92/24 = 0.0384).
12. (a) The SPECT myocardial perfusion study tries to evaluate initial tracer distribution as an indication of coronary circulation. Any tracer redistribution (more prominent for ^{201}Tl, minimal for sestamibi) has only adverse effect on imaging quality.

13. (c) Dipyridamole, adenosine, and dobutamine can be used to produce pharmacologic stress. The side effects of dipyridamole can be reversed by using aminophylline, which should be on hand during, and for 1 h following, dipyridamole administration. Side effects of dipyridamole include angina, dizziness, and headaches and nausea.
14. (e) The R wave signals the start of data collection, but each R-R interval is divided into segments, so if the R-R interval varies, as it does in arrhythmia, the data will be degraded. If this data is filtered out, the exam time will be increased.
15. (b) Vertical long-axis slices show sections of the left ventricle from the septum to the free wall and are most similar to sagittal images.
16. (b) Aminophylline may be used to reverse the side effects of dipyridamole, which is used to create pharmacologic stress, as are adenosine and dobutamine.
17. (e) End systole is represented by the frame with the lowest number of counts in the ROI around left ventricle.
18. (b) The quality of a non-gated study would not be affected by arrhythmia, but using the wrong collimator or center of rotation would, as would increased patient-to-detector distance.
19. (a) The patient's weight must first be converted from pounds to kilograms (155 lb. divided by 2.2 lb./kg = 70.5) and then multiplied by 0.56 mg to find the required dose (39.5 mg required). Since each mL of dipyridamole contains 5 mg, 39.5 is divided by 5 to find the number of mL to inject (39.5/5 = 7.9).
20. (e) Patients should be on NPO for at least 4 h before the exam, except for water for medicine. Beta-blocker should be avoided for exercise stress test. Caffeinated food and beverages are not allowed for chemical stress test but are not restricted for treadmill test.
21. (c) In the situation described, the most likely explanation is a disconnected ECG lead and can quickly be confirmed. If this is not the case, medical assistance should be immediately sought.
22. (d) ^{99m}Tc sestamibi is used for myocardial perfusion imaging, ^{99m}Tc pyrophosphate was used for myocardial infarct detec-

tion and currently used for evaluation of cardiac amyloidosis, and ^{123}I MIBG can be used for heart failure evaluation. ^{32}P is a beta emitter and has therapeutic applications.

23. (e) All of the above.
24. (a) Adenosine has a half-life of only ~10 s.
25. (a) Including the spleen within the ROI for the background will give a higher background activity, and thus higher EF.
26. (a) Rubidium-82 is the daughter of ^{82}Sr.
27. (a) The action potential that causes contraction arises in the sinoatrial (SA) node, which is located in the right atria. It then travels to the AV node and through the bundle of His. The bundle of His branches into the right and left bundle branches, which end in the Purkinje fibers; these innervate the individual contractile myocardial cells. The P wave is that part of the ECG that represents the depolarization of the atria.
28. (e) A commonly used exercise protocol is the Bruce protocol, which gradually increases both the speed and the grade of the treadmill. A false-negative exercise myocardial perfusion study may be the result of insufficient exercise.
29. (b) The high photon flux involved with ^{99m}Tc sestamibi necessitates the use of a high-resolution collimator. It is sufficient to use a low-energy all-purpose collimator (LEAP) for the ^{201}Tl exam (as well as the infarct study with pyrophosphate). High-sensitivity collimation is preferable to a first-pass study.
30. (b) The half-life of ^{82}Rb is 75 s.
31. (b) ^{82}Rb is generator produced. Given the short half-life of rubidium, a generator must be on-site in order for a study to be performed utilizing ^{82}Rb.
32. (a) The number of views is multiplied by seconds per view and divided by 60 to obtain the total number of minutes, and then divided by 2 for dual-head camera, which is most commonly used.
33. (b) Gated blood pool imaging of the heart can be analyzed to determine the left ventricular ejection fraction. Hibernating myocardium and ischemia can be assessed using myocardial perfusion imaging, and intracardiac shunt may be determined by a gated first-pass study.

34. (a) Akinesis means no wall motion, dyskinesis means paradoxical wall motion, and hypokinesis means diminished wall motion.
35. (c) The R-R interval represents the length of the cardiac cycle. In patients with rhythm disturbances, the R-R interval may vary. Since gated imaging involves division of the interval into equal parts, if the interval is not of constant length, the part of the cycle that is being imaged will vary from cycle to cycle, and the resulting images will be compromised. For this reason, arrhythmia filtering is used; data from cardiac cycles that vary too much in length are rejected.
36. (b) The highest labeling efficiency results from in vitro methods, although the other methods are still considered acceptable for use. The in vivo method results in free technetium localizing in the kidneys, bladder, stomach, thyroid, and salivary glands.
37. (c) The usual dose of Rb–Cl is 30–60 mCi for each of the stress and rest scans. This is dependent on camera capabilities. Too high a dose may flood the crystal, making imaging impossible.
38. (c) Infarcted tissue does not perfuse; therefore, you would see areas of low or no uptake.
39. (d) Gated blood pool ventriculography is useful for determining the effect of chemotherapy on cardiac function. In general, if the LVEF is less than 45%, the cardiotoxic therapy may be discontinued, or if a change in LVEF is more than 15%, the therapy may be considered to be too dangerous to cardiac function to continue.
40. (b) In the case of myocardial perfusion imaging, the separation of the left atrium from the left ventricle is not as important as it is in ventriculography, so the caudal tilt is not important. All of the other items are important for SPECT myocardial imaging with ^{99m}Tc sestamibi.
41. (c) Dipyridamole, adenosine, and dobutamine may all be used to produce pharmacologic stress or as an adjunct to exercise. All of these are associated with side effects, but both dipyridamole and adenosine may cause or worsen broncho-

spasm and so should not be used in patients with asthma or COPD.

42. (a) ^{82}Rb PET can be performed before FDG-PET in the same day, due to its very short half-life.
43. (d) ^{99m}Tc has superior imaging characteristics compared to ^{201}Tl and can be used to image myocardial perfusion at 1–3 h after injection. The disadvantage to ^{99m}Tc sestamibi is that two doses need to be injected for imaging of perfusion under stress and at rest. Another alternative is to perform a resting study with ^{201}Tl followed by a stress study using ^{99m}Tc sestamibi. In addition, a first-pass study and gated myocardial images can be obtained with ^{99m}Tc sestamibi, although radionuclide ventriculography is typically done with labeled erythrocytes or albumin. ^{99m}Tc PYP and labeled erythrocytes are not used for the assessment of myocardial perfusion.
44. (a) The tracer first enters the right heart, leaves the right heart via the pulmonary artery, and enters the lungs. The oxygenated blood enters the pulmonary vein and is carried to the left side of the heart.
45. (d) Attenuation from breast tissue or breast implants may create the appearance of defects in the anterior and anteroseptal walls of the heart.
46. (b) The lower dose is always administered first, no matter if the rest or stress is performed first.
47. (b) The second injection of ^{99m}Tc sestamibi is required and with higher dose, because unlike ^{201}Tl, there is no redistribution of sestamibi in the myocardium.
48. (d) SPECT myocardial imaging with ^{201}Tl chloride tends to have more attenuation artifacts because of the lower photon energy of ^{201}Tl.
49. (b) A bull's-eye display or polar map is created from the entire set (either stress or rest) of perfusion images, so it includes all walls of the myocardium. It offers a convenient way to compare stress and rest datasets.
50. (c) ^{99m}Tc sestamibi has more hepatic and GI activity with ^{201}Tl, so (a) is not correct. There is less soft tissue attenuation with ^{99m}Tc, so (b) is incorrect. The delay between exercise

and imaging is shorter for ^{201}Tl, so (d) is also incorrect. There is a higher photon flux from the ^{99m}Tc dose.

51. (d) ^{99m}Tc sulfur colloid is used to evaluate lower extremity lymphedema, typically injected intradermally in each foot.
52. (b) Initial uptake of ^{99m}Tc sestamibi into the liver and lungs can obscure the myocardium. For this reason, imaging does not begin immediately. A fatty meal or milk may stimulate gallbladder contraction, and some activity may be distributed into the bowel.
53. (a) The lower energy isotope (^{201}Tl) should be imaged first, to prevent interference from the higher energy photons. Because attenuation is a bigger problem with ^{201}Tl, the exercise study is best performed with ^{99m}Tc.
54. (d) ^{82}Rb cardiac perfusion study cannot determine myocardial viability.
55. (c) If the patient fails to reach maximal stress, false negatives may result. The maximal stress is usually defined as 85% of a predicted maximum heart rate (usually 220 − age in years) but may also be the point that chest pain or significant ECG changes occur, or when the systolic blood pressure multiplied by the heart rate is greater than 25,000.
56. (d) The dipyridamole (0.56 mg/kg body weight) is injected first, normally over 4 min. The radiopharmaceutical is injected after 7 or 8 min, and the aminophylline should be prepared in advance for injection in case of serious side effects (side effects may include chest pain, dizziness, headaches, and nausea).
57. (d) The data collection time may be fixed by setting a count number to be reached or by setting the number of cardiac cycles to image.
58. (b) In vivo means that the labeling takes place inside an organism. In vitro means that the labeling takes place outside of the organism. The modified in vivo procedure involves injecting the reducing agent into the patient and then withdrawing and labeling the blood in vitro before reinjecting the blood.
59. (f) (a) and (c) only. Myocardial viability does not need the stress part.

60. (d) See explanation for Question 57. Depending on what parameters are set for data collection (total counts or cardiac cycles), all of the choices may affect imaging time.
61. (e) MUGA is used to evaluate LV ejection fraction and wall motion/contraction, not myocardial ischemia.
62. (b) A good bolus injection is not required (no dynamic imaging during injection is needed) because this study is to evaluate the blood pool activity (thus imaging is obtained 20 min or later after tracer injection).
63. (d) Treadmill exercise test can still be performed if the patient had coffee or tea before the study.
64. (c) Coffee and tea are contraindicated in stress test when using Persantine, Adenosine, or Lexiscan. Coffee and tea are stimulant agents to the central nervous system and should be avoided for FDG-PET study of the brain.
65. (c) Aminophylline is used to treat patient's side effects from Persantine or Lexiscan and should be used 2 min or more after tracer injection.
66. (e) Sestamibi should be injected during the last minute of dobutamine administration, immediately after Lexiscan administration, or 2 min after Persantine administration.
67. (c) ^{99m}Tc-PYP is used for the evaluation of suspected transthyretin cardiac amyloidosis (also known as ATTR cardiac amyloidosis). SPECT imaging is preferred.
68. (b) For thallium study, stress test has to be performed first. For sestamibi study, either rest or stress test can be performed first.

Appendix 9: Answers to Chapter 9

27

1. (b) A typical dose of MAA for lung perfusion study has 200,000–600,000 particles, blocking less than 1 in 1000 capillaries (<0.1%).
2. (d) When injecting a patient for a perfusion study, he or she should assume a supine position to ensure uniform distribution of particles. If blood is withdrawn into the syringe to confirm venous access, the radiopharmaceutical should be injected immediately as aggregated albumin and blood will form clots if left to sit. Intravenous lines or indwelling catheters that have filters should also be avoided to ensure that the particles reach the lung capillaries.
3. (b) The amount of technetium added to the kit will not affect the distribution of the aerosol particles, but only a small portion of the activity administered will actually be delivered to the patient. (It is estimated that of 25–35 mCi added to the nebulizer, a patient will receive 1–5 mCi. Therefore, it is important to do the ventilation scan first when planning both ventilation with aerosol and perfusion imaging.) The distribution of the particles may be altered if the patient has COPD (which impedes flow rate and causes turbulence) or if the aerosol particle size is large. Breathing the aerosol in the supine position will improve uniformity of distribution.
4. (d) The possibility to study the single breath, equilibrium, and washout phases is an advantage of ^{133}Xe ventilation imaging.

E. Mantel et al., *Nuclear Medicine Technology*,
https://doi.org/10.1007/978-3-031-26720-8_27

During the single-breath phase, the patient exhales as completely as possible and then inhales ^{133}Xe and holds it while an image is taken. During equilibrium, the patient breathes a mixture of ^{133}Xe and oxygen. During the washout phase, fresh air is inhaled, and ^{133}Xe is exhaled from the lungs.

5. (a) Patients should be injected for perfusion scans while supine to ensure uniform distribution to the lungs.
6. (b) See explanation to Question 5.
7. (d) Normally, the first capillary bed that the radiopharmaceutical encounters is in the lungs. If there is a right-to-left cardiac shunt, renal activity will be present, and the head should be imaged to confirm.
8. (d) The method is mechanical, based on the size of the particles in the radiopharmaceutical being too large to pass through the capillaries.
9. (e) Clots formed when albumin is allowed to contact sitting blood will create hot spots on perfusion images. This can happen if the radiopharmaceutical is not promptly injected after blood was withdrawn into the syringe.
10. (a) The trachea branches to two bronchi, which in turn branch into the bronchioles, the smallest of which branch into the alveoli.
11. (d) See explanation to Question 3.
12. (c) Activity in the central airway (often due to turbulent flow in patients with COPD) and stomach (from swallowed droplets) is often present on ventilation scans using aerosolized radiopharmaceuticals.
13. (d) It is preferable to inject directly into the vein (if an indwelling catheter is used, it should be flushed following injection), and 800,000 particles are beyond the recommended dose. Supine positioning during injection increases the uniformity of particle distribution.
14. (d) ^{81m}Kr is useful for ventilation imaging. Its short half-life (13 s) and slightly higher photon energy allow the technologist to perform ventilation before or after perfusion imaging or, as a simultaneous scan, positioning once, imaging one isotope after another before moving the patient to the next position. The short half-life of ^{81m}Kr (13 s) may result in less

exposure for technologists (in the event that the patient removes the ventilation mask) than with ^{133}Xe.

15. (b) Since most particles will be cleared by the lung capillaries on the first pass, no wait is needed.
16. (d) Risk factors for pulmonary embolism include oral contraceptive use or hormone therapy, prolonged inactivity (e.g., bed rest), recent surgery, and being postpartum. Some, but not all, sources cite smoking and obesity as risk factors.
17. (c) Those who should be considered for lower particle doses are the young, those with pulmonary hypertension, or those with right-to-left shunting.
18. (d) The dose administered with aerosolized ^{99m}Tc DTPA will be much higher than the dose the patient actually receives. The Society of Nuclear Medicine Guidelines state that of 25–35 mCi administered, the patient will receive only 0.5–1 mCi.
19. (b) Lung quantitation may be used preoperatively in patients for whom lung resection is planned.
20. (c) The total counts are 197,980. Those from the right lower lobe are 41,502; this number is divided by total counts and multiplied by 100 to find 20.9%.
21. (b) A mismatch, that is, a perfusion defect in the absence of a matching ventilation defect, often indicates pulmonary embolism.
22. (d) The number of particles in 2 mL will be half of that in the kit of 4 mL.
23. (c) The specific concentration is 17.5 mCi/mL. To administer 4 mCi, 0.23 mL would have to be injected (activity required/specific concentration). Since there are 1,250,000 particles per mL, the patient will receive 287,500 particles (particle concentration multiplied by mL injected).
24. (a) The chest X-ray may be normal or nonspecific in pulmonary embolism but may show other diseases that may aid interpretation of the lung scan. In any case, a recent chest X-ray should be performed before the scan.
25. (a) At 2 p.m., the activity has decreased by one half-life, meaning that the activity is 45 mCi, and there are 8,000,000 particles in 5 mL provided that no dose has been withdrawn

from the vial. The concentration is now 1,600,000 particles per mL and 9 mCi per mL. If 0.4 mL is withdrawn into a syringe, there will be 3.6 mCi and 640,000 particles in the dose.

26. (b) The number of particles will not change because the particles and the volume remain constant, but the number of particles per mCi will change since the activity is decaying.
27. (c) Total lung counts are 338,103. Right lung counts divided by total counts gives a percentage of 57.7%.
28. (c) The capillaries surrounding the alveoli measure 0.7–10 μm in adults. MAA particles measure 5–100 μm; most are between 10 and 30 μm. Particles greater than 10 μm are trapped.
29. (c) Particle fragments break down (some as soon as 30 min after injection), enter the bloodstream, and then are removed by the liver and spleen.
30. (c) Ventilation images can be performed with the patient erect or supine. Posterior images reduce patient-to-detector distance and avoid breast attenuation, particularly important with lower energy photon like that from ^{133}Xe.
31. (a) A nebulizer is used to create a mist from a liquid and is used for ventilation scanning without gas. As krypton and xenon are true gases, a nebulizer is not needed.
32. (b) See explanation for Question 24.
33. (a) ^{133}Xe has a lower photon energy (81 keV) than ^{99m}Tc (140 keV), and ventilation scans are usually performed first when using this gas.
34. (c) Since ^{133}Xe is heavier than air, it will drop and can be trapped by a floor system. In addition, ventilation scans should be performed in a room at negative pressure to surrounding areas.
35. (d) The stomach is often seen on aerosol lung scans for this reason. See explanation for Question 12.
36. (b) Typically, 25–35 mCi ^{99m}Tc DTPA is administered; however, the patient will receive only 0.5–1 mCi.
37. (b) Much less than half is delivered. For this reason, aerosol ventilation scans are usually performed before perfusion scans.

38. (e) A patient must understand the effect of cooperation on the quality of the exam and of the increased exposure to healthcare workers if the ventilation apparatus is removed prematurely.
39. (b) Xenon is fat soluble and partially soluble in blood and often appears during the washout phase of the ventilation scan.
40. (c) The ventilation can be performed after the perfusion scan because of the higher photon energy of ^{127}Xe relative to ^{99m}Tc (203 keV vs. 140 keV), and, therefore, the ventilation scan may be canceled if the perfusion is negative. It has a 36-day half-life and is expensive and is therefore not often used.
41. (d) Because ^{81m}Kr has a very short half-life (only 13 s), and because ^{81m}Kr has higher energy photon emissions compared with ^{99m}Tc-MAA, ^{81m}Kr ventilation studies can be performed either before or after perfusion imaging and can be repeated as needed.
42. (b) See explanation for Question 35. It is not likely that the safe conditions will be present in the ICU.
43. (e) In addition, a V/Q scan can be used to evaluate hepatic pulmonary syndrome and quantitatively evaluate lung perfusion for lung transplants.
44. (e) The dose used for the ventilation imaging is about 30 mCi ^{99m}Tc DTPA, but only ~1 mCi is delivered to the lungs. ^{99m}Tc MAA dose is ~5 mCi and thus has to be performed after ventilation.
45. (d) In most cases, the number of particles should be in the range of 200,000–700,000 in adults, but for patients with pulmonary hypertension or right-to-left shunting, the number may be reduced to 100,000–200,000 particles in adults. All other answers are correct.
46. (c) One of the major disadvantages of aerosol ventilation imaging is that aerosol deposition is altered by turbulent flow, causing central deposition and suboptimal imaging.
47. (c) In fact, CTA has more incidental findings to explain patient's symptoms.
48. (d) Tracer uptake in the thyroid, stomach, and kidneys could be due to free TcO_4^-.

49. (c) Ventilation imaging should not be performed in patients with confirmed or suspected Covid-19 infection, due to increased risk of aerosol production, contamination, and leakage of the aerosol from the closed delivery system in the imaging room. Perfusion-only study should be performed instead, ideally with SPECT with low-dose CT.

Appendix 10: Answers to Chapter 10

28

1. (c) To simultaneously image gastric emptying of liquids and solids, two radionuclides are necessary, so choice (b) is incorrect. Choice (a) contains two solids and so is incorrect. The most common practice is to use ^{99m}Tc sulfur colloid with whatever solid meal is preferred by the given laboratory along with ^{111}In in liquid.
2. (b) Figure 10.1 shows the stomach outlined by regions of interest for the purpose of calculating the percentage of emptying over time.
3. (d) Damaged red blood cells are removed from circulation largely by the spleen, so tagged damaged erythrocytes as well as tagged colloids can be used to image the spleen.
4. (b) A LeVeen shunt routes ascites from the peritoneum to the superior vena cava; therefore, the radioisotope is injected into the peritoneal space.
5. (c) Cholecystokinin, or CCK, is a hormone secreted by the duodenum, which causes the gallbladder to contract and secrete bile. Sincalide is a synthetic form of CCK that can be used before radiopharmaceutical injection in patients who have fasted for extended periods of time or who receive parental nutrition to empty the gallbladder, allowing the radiopharmaceutical to localize there. When used, it should be injected over 5–10 min to ensure that spasm is not induced.

E. Mantel et al., *Nuclear Medicine Technology*,
https://doi.org/10.1007/978-3-031-26720-8_28

It can also be used to determine gallbladder ejection fraction (normal is >50%).

6. (c) Patients should fast 2–4 h before hepatobiliary imaging. The gallbladder may not be visualized in patients who have recently eaten and in patients who have fasted for extended periods. The use of morphine can hasten visualization of the gallbladder by increasing bile flow to the gallbladder relative to the intestine. For patients in an extended fast, sincalide may be used (see explanation to Question 5). If there is no gallbladder seen 1 h following injection of radiopharmaceutical, delayed images should be obtained.
7. (b) Lung activity on a liver/spleen scan using ^{99m}Tc sulfur colloid may be the result of clumping of particles, a particle that is too large or too much Al^{3+} ion in the technetium used to prepare the kit.
8. (c) See explanation for Question 4.
9. (b) The technologist who sees that the exam ordered is inappropriate for addressing the question asked by the referring physician should alert the departmental physician in charge so that the situation can be resolved. It is irresponsible to perform the exam as ordered, but beyond the responsibility of the technologist to change the order.
10. (d) The camera rotation should be as close as possible to the patient without touching him or her. Any IV poles, catheter bags, or other attachments should be carefully placed to avoid entanglement during rotation. The detector head should be level, and the arms should ideally be positioned above the head.
11. (c) Radiocolloids are cleared largely by the liver's Kupffer cells but also by the spleen and the bone marrow. The smallest particles tend to be sequestered by the bone marrow, and the largest by the spleen. In certain hepatic conditions, colloid shift may be present; this manifests as greater clearance by the spleen and bone marrow relative to the liver.
12. (b) Cavernous hemangioma is a common vascular malformation which often causes no symptoms in patients. It can be evaluated using ^{99m}Tc-labeled red blood cells. Increased activity is seen on initial perfusion images, which decreases

on delayed static images. Large hemangiomas may show inhomogeneous uptake due to thromboses. SPECT has a much greater sensitivity for detecting cavernous hemangioma than does planar imaging.

13. (c) See explanation for Question 11.
14. (d) The listed choices (a), (b), and (c) can all be performed using labeled erythrocytes. ^{99m}Tc pertechnetate is taken up by gastric mucosa. Meckel's diverticulum is a congenital outpouching of the intestine, which may contain ectopic gastric mucosa, so imaging with ^{99m}Tc pertechnetate can be useful for diagnosis.
15. (b) Cimetidine inhibits release of pertechnetate from the gastric mucosa. It may be administered to patients 1–2 days prior to performing an exam to improve sensitivity of detection of Meckel's diverticulum.
16. (c) The half time for solid gastric emptying is dependent on the type of meal used but is about 90 min for scrambled eggs, a commonly used meal.
17. (b) Esophageal reflux studies typically use a radiopharmaceutical in liquid.
18. (d) Applying an abdominal binder during an esophageal reflux study helps to increase abdominal pressure over time while imaging and thus increase the sensitivity of the study.
19. (a) Only ^{99m}Tc pertechnetate is used for the diagnosis of Meckel's diverticulum.
20. (c) The uptake and distribution following injection of a hepatobiliary imaging agent depend in part on the patient's bilirubin level. Increased level of bilirubin is associated with decreased excretion of the radiopharmaceutical.
21. (b) Iminodiacetic acids (IDAs) are cleared from the blood by the hepatocytes before they are secreted into the canaliculi. If the liver is not seen but there is cardiac and renal activity at 15 min after injection, this likely indicates that there is a problem of liver function.
22. (d) See explanation for Question 5.
23. (e) If there is insufficient fasting, radiopharmaceutical may not be able to enter the gallbladder, resulting in a false positive.

24. (b) Images pre- and post-CCK or fatty meal are used to draw regions of interest for calculation of gallbladder ejection fraction. The formula used to calculate gallbladder ejection fraction is (net maximum counts – net minimum counts/net maximum counts) × 100%. Since there is no mention of background counts in the question, one assumes that the numbers given are in net counts. ((285,000 – 187,000/285,000) × 100 % = 34%.)
25. (b) Normal gallbladder ejection fraction is >35%.
26. (d) Within an hour, a normal hepatobiliary scan (e.g., with 8 mCi ^{99m}Tc DISIDA) is expected to show the liver, the common bile duct, the gallbladder, and the small intestine.
27. (a) Cholecystokinin is secreted by enteroendocrine cells in the duodenum and stimulates the contraction of the gallbladder and the relaxation of the sphincter of Oddi.
28. (d) As morphine enhances the muscle tone of the sphincter of Oddi, pressure will increase in the bile ducts and may result in gallbladder filling. This may be an alternative to delayed imaging.
29. (b) ^{99m}Tc sulfur colloid is rapidly cleared from the bloodstream, and the patient must be actively bleeding while sulfur colloid is intravascular for the bleed to be visualized. Delayed imaging is therefore not possible, but repeat studies can be undertaken within a relatively short time. Increasing liver and spleen activity may complicate the detection of bleeding in these areas.
30. (e) The use of cine mode, oblique imaging, and delayed imaging is useful in the detection of GI bleeds.
31. (a) Potassium perchlorate may be given to reduce thyroid uptake of pertechnetate or radioiodine. However, it should not be given to patients undergoing scanning for Meckel's diverticulum with ^{99m}Tc pertechnetate, because it also blocks secretions from gastric mucosa and may result in a false-negative scan.
32. (a) Labeling the cells in vitro will result in the highest labeling efficiency, i.e., the least amount of pertechnetate.
33. (d) Free technetium will cause increased activity in the kidneys and gastric mucosa and passes into the bladder, small

bowel, and colon and thus may interfere with image interpretation.

34. (b) Newer guidelines recommend images be obtained at hour intervals, up to 4 h, either erect or supine (depending on patient condition).
35. (e) No fasting is necessary before scanning to rule out a GI bleed; some fasting is required for all of the other exams listed.
36. (d) Symptoms of both acute and chronic cholecystitides include nausea and pain, which can be limited to the abdomen but sometimes extends to the back and right scapular area.
37. (b) Bile is produced in the liver and then enters the biliary canaliculi. From there, approximately two-thirds of bile bypasses the gallbladder and enters the duodenum. The remaining bile is stored in the gallbladder until it is needed and secreted.
38. (c) Gastric emptying study should be terminated if the patient vomits and is finished if less than 10% of the meal remains in the stomach.
39. (d) Often, ^{99m}Tc is attached to sulfur colloid and bound to egg and used for a solid-phase study, while ^{111}In-DTPA or ^{99m}Tc-DTPA is used for a liquid-phase study.
40. (e) GI bleeding imaging is normally for 60 min but can be longer. If more imaging is needed, it should be performed continuously. Delayed imaging after interruption often does not help.
41. (e) Other medicines that affect ^{99m}Tc RBC labeling include the following: methyldopa, hydralazine, quinidine, etc.
42. (d) All can be used, but cimetidine is most commonly used.
43. (d) For HIDA scan, patients need to be NPO for at least 4 h but not longer than 24 h, and opioid narcotics should be held for at least 4 h. Caffeine does not interfere with this test.
44. (c) Meckel scintigraphy should be used when the patient is not actively bleeding. For patients with active bleeding, radiolabeled red blood cell (RBC) scintigraphy should be performed.
45. (c) A 24-h delayed imaging is often required, unless bowel activity is seen earlier.

46. (d) Patients should discontinue medications that affect motility before the study especially for medications that are not administered on a daily basis. Medications such as opiate analgesics slow gastrointestinal transit, while prokinetic agents (such as metoclopramide and erythromycin) accelerate gastrointestinal transit.
47. (b) Scintigraphic methods for measuring small-bowel transit have been in use for at least 20 years but have not gained widespread application. Different protocols exist for the evaluation of small-bowel transit. ^{99m}Tc attached to sulfur colloid and bound to egg or ^{111}In-DTPA is commonly used for bowel motility studies and is nonabsorbable in the bowel.

29 Appendix 11: Answers to Chapter 11

1. (b) Because of the rapid change in appearance of the kidney in early images, one can assume that these images are of very short duration, ruling out choices (c) and (d). Because of the activity in the kidney in the last image, and the appearance of arterial flow on just one image, one can assume that they are 3-s images taken over 45 s. The arterial phase would be more clearly seen on multiple images if they were 1-s images.
2. (a) On delayed images, there is often decreased or even absent activity in early-phase torsion. Perfusion may show normal or decreased activity unless the torsion is late phase. In this case, perfusion is increased in the tissue surrounding the testicle.
3. (b) ^{99m}Tc MAG3 is cleared by the proximal tubules. ^{99m}Tc DTPA is cleared by glomerular filtration, which makes it useful for estimating glomerular filtration rate (GFR). ^{99m}Tc DMSA binds to proximal renal tubules, is useful for imaging the renal parenchyma, and clears very slowly in the urine. ^{99m}Tc GH is cleared by both glomerular filtration rate and renal tubules and is also useful for imaging the renal cortex.
4. (d) Injected ^{99m}Tc DMSA will remain concentrated in the renal cortex for many hours after injection. Delayed imaging is often necessary to have sufficient activity in the kidneys relative to background if the patient has impaired renal function.

E. Mantel et al., *Nuclear Medicine Technology*, https://doi.org/10.1007/978-3-031-26720-8_29

5. (d) MIBG is taken up by chromaffin cells, which are in the adrenal medulla and the paraganglia of the sympathetic nervous system; it is usually labeled with ^{123}I or ^{131}I and is used to image neuroectodermally derived tumors such as pheochromocytomas, neuroblastomas, and paragangliomas. ^{68}Ga DOTATATE is a newer PET tracer to evaluate neuroendocrine tumors including pheochromocytomas.
6. (c) Lugol's solution will block some of the uptake of ^{131}I by the thyroid.
7. (d) ^{99m}Tc DMSA may need pinhole cortical images or SPECT images. Typically, dynamic imaging is obtained for ^{99m}Tc DTPA and ^{99m}Tc MAG3.
8. (b) Although the dose can be lower than other ^{99m}Tc renal imaging agents and the photon is not high energy, the radiation dose to the kidney is higher than with other agents because of persistent DMSA activity in the kidneys for many hours.
9. (a) See explanation for Question 3.
10. (d) Although CT and MR can evaluate renal anatomy, radionuclide imaging is still useful for all of the indications listed.
11. (d) Glomerular filtration is the process that removes excess water, salts, and urea from the blood. This filtration takes place in the glomeruli following which the filtered blood flows into Bowman's capsule. GFR varies according to age, race, and gender, but the approximate normal value is 120–125 mL/min for both kidneys in young and middle-aged adults.
12. (a) See explanation for Question 3.
13. (b) Patients undergoing renal imaging with ^{99m}Tc MAG3 should be well hydrated and should void just before the exam. The renogram may show delayed peak activity and delayed clearance of radiopharmaceutical if the patient is dehydrated.
14. (d) If an obstruction is suspected, a diuretic can be administered, the patient should be asked to void, or be catheterized, and further imaging may be carried out.
15. (d) Diuretic renography, usually performed using ^{99m}Tc MAG3 with furosemide, can help differentiate between functional abnormalities and anatomic obstruction. Furosemide

will increase the pressure in the renal pelvis causing the tracer to move from the collecting system, unless the obstruction is anatomic; in this case, there would be little change in the distribution of the tracer.

16. (b) Imaging of transplants is performed anteriorly with the detector centered over the relevant iliac fossa.
17. (b) The right kidney is often, but not always, lower than the left, probably because of its proximity to the liver.
18. (d) Diuretic renography is used to evaluate collecting system obstruction.
19. (c) Testicular torsion is a painful condition in which the spermatic cord is twisted, and it usually requires surgical correction within 5–6 h. It often follows trauma, but may also be caused by a congenital abnormality, or may be spontaneous. The involved testicle is usually swollen.
20. (d) Dosage varies from 5 to 30 mCi; 20 mCi is a commonly recommended dose.
21. (b) The blood supply to the native kidneys comes from the renal arteries, which arise from the descending aorta.
22. (c) Radionuclide cystography is performed to evaluate patients for vesicoureteral reflux. The radiopharmaceutical is introduced into the bladder with saline, and the patient is imaged during bladder filling and voiding to detect reflux of activity from the bladder into the ureters.
23. (c) A formula for estimating bladder capacity is needed to prepare an appropriate amount of saline for bladder filling. The formula used for >1-year-old children is as follows: $(\text{Age} + 2) \times 30$ mL.
24. (c) See explanation for Question 23. Quantitative information can be obtained by comparing counts in the bladder pre- and post-voiding and by using regions of interest in the ureters.
25. (d) The possibility to image any reflux will not be lost, but there is danger that the detector and other equipment will become contaminated, and some quantitative information may be unreliable.
26. (a) Normal cystography will show an increase in activity in the bladder during filling and decreasing activity during voiding with no activity in the ureters or kidneys.

27. (e) Expected bladder capacity must be calculated (see explanation for Question 23). It is important to ascertain the patient's ability to void on request, and the detector head and surrounding equipment should be protected against contamination from the radioactive urine. The patient should be catheterized, and the bladder should be emptied before filling.
28. (b) Any reflux into the ureters or kidneys is abnormal. Small volumes of reflux into the distal ureters may be difficult to detect.
29. (a) Imaging should be dynamic during filling and voiding.
30. (c) (See explanation for Question 11.) Glomerular filtration rate is the amount of fluid filtered from the glomerular capillaries to the Bowman's capsule in a given amount of time, normally expressed as mL/min. It is defined as the ability of the kidneys to clear inulin from plasma; inulin is a substance that is filtered by the glomeruli and is not secreted or reabsorbed by the tubules. Significant kidney function may be lost before serum creatinine becomes abnormal, so GFR is an important measurement. ^{99m}Tc DTPA is cleared by the glomeruli and is used for measuring glomerular filtration rate. Camera-based methods require no blood sampling.
31. (d) Renal plasma flow is defined as the clearance of para-aminohippurate (PAH, which is secreted and filtered by the renal tubules) from plasma. In nuclear medicine, effective renal plasma flow (ERPF) can be measured using ^{131}I OIH or ^{99m}Tc MAG3.
32. (a) Blood tests with implications for renal function, like BUN and creatinine, only capture information about the overall renal function. Quantitative studies of GFR and ERPF can provide information about individual kidney function.
33. (c) See explanation for Question 23.
34. (e) ^{99m}Tc DMSA is usually administered in lower doses because of its retention in the kidney. ^{131}I OIH, with its beta emissions and long half-life, also has a lower dosage (but OIH may also be tagged with ^{123}I). ^{99m}Tc MAG3 and ^{99m}Tc DTPA have similar dosages.
35. (b) DMSA should be discarded after 4 h.
36. (d) A small percentage of ^{99m}Tc MAG3 may be taken up in the liver and gallbladder in normal patients, and this may be more pronounced in patients with poor renal function.

37. (d) Dynamic imaging is obtained for ^{99m}Tc DTPA and ^{99m}Tc MAG3. ^{99m}Tc DMSA has slow background clearance, and only delayed cortical planar or SPECT imaging is needed.
38. (a) The adrenal glands sit atop the kidneys.
39. (b) The glomerulus is a tuft of capillaries within the nephron. The loop of Henle is the hairpin loop that the tubule makes. Distal tubules from several nephrons join to form the collecting duct. The renal pyramids are part of the renal medulla, which empty into the renal pelvis.
40. (b) About 25% of the cardiac output is directed to the kidneys.
41. (d) If the images are obtained posteriorly, the lower poles may appear to have slightly less activity than the upper poles because the lower poles are often situated slightly anterior to the upper poles.
42. (b) Patients who have renal artery stenosis, when given an ACE (angiotensin-converting enzyme) inhibitor, will show a decrease in GFR. Therefore, captopril renography is an effective examination to determine whether renal artery stenosis is the cause of hypertension.
43. (a) In captopril renography, the captopril is administered orally 1 h before the injection of the radiopharmaceutical.
44. (d) The patient should be well hydrated, fasting, and off ACE inhibitors before the study is begun. The length of time that the medications should be held depends on the type of medication, typically 3 days for captopril and 1 week for some other medications.
45. (b) Captopril renography is used when the patient is suspected of having renovascular hypertension, but since only 1–4% of hypertension is due to renovascular hypertension, it is not used as a screening test in all cases of hypertension.
46. (d) The renogram shown indicates normal renal function with no evidence of obstruction. No additional images needed.
47. (d) If a patient is already on ACEI therapy, ACEI should be discontinued for 3–5 days (depending on half-life) before ACEI renography.

Appendix 12: Answers to Chapter 12

30

1. (d) ^{99m}Tc sestamibi, ^{201}Tl chloride, and ^{67}Ga citrate are nonspecific tumor markers and had been historically used for tumor imaging. However, they have been largely replaced by newer PET tracers.
2. (d) ^{67}Ga has a half-life of 72 h, and so imaging can take place over several days. Imaging for infection is often performed on the same day as the injection and continues at 24 h and onward at 24-h intervals. Tumor imaging is usually begun at 48 h and continued at 24-h intervals.
3. (d) Lymphoscintigraphy is performed by injecting a radiocolloid near a tumor and is used to identify the sentinel node(s).
4. (c) Pentetreotide (octreotide) is a synthetic form of somatostatin. Somatostatin is a neuropeptide that, among other things, inhibits the secretion of growth hormone. Somatostatin receptor imaging is performed for the detection of neuroendocrine and some non-neuroendocrine tumors with high density of somatostatin receptors. These are often tumors that originate from somatostatin target tissue. Examples of cancers that express somatostatin receptors are pancreatic islet cell tumors, certain brain tumors, pituitary adenoma, gastrinoma, small-cell lung cancer, pheochromocytoma, medullary thyroid carcinoma, some breast cancers, and lymphomas.

E. Mantel et al., *Nuclear Medicine Technology*,
https://doi.org/10.1007/978-3-031-26720-8_30

5. (b) HAMA stands for human antimurine (aka antimouse) antibody. It is a response that the human immune system may mount against a foreign antigen.
6. (a) Different injection methods have been used for lymphoscintigraphy, including deep (subcutaneous, peritumoral, and intratumoral) and superficial (intradermal and subdermal) injection. Generally, intradermal injection is preferred.
7. (d) Sentinel node lymphoscintigraphy is performed to identify potential metastatic lymph nodes. A breast cancer patient with known metastasis (including palpable axillary nodes) should not have a lymphoscintigraphy.
8. (c) Sentinel node lymphoscintigraphy in a patient with melanoma in the body trunk should include bilateral axillae and groins.
9. (a) ^{99m}Tc sestamibi, most commonly used in myocardial perfusion study, is a nonspecific tumor marker and is also commonly used in the evaluation of parathyroid adenoma.
10. (b) Antibodies are proteins that are produced in response to the presence of an antigen, and they bind to that antigen. Cross-reactivity means a reaction to a similar but different antigen.
11. (d) Neuroendocrine tumors have been called apudomas (taking the first letters from amine precursor uptake and decarboxylation) or tumors arising from APUD cells. (APUD cells secrete polypeptides, amines, or both.) The neuroendocrine system compromises the nervous system and the endocrine glands and the interactions between the two. Examples include carcinoids, small-cell lung cancer, pituitary adenomas, pheochromocytomas, some brain tumors, and islet cell tumors.
12. (a) Genitourinary activity may be seen even on delayed images obtained with ^{111}In pentetreotide. It is not normally seen on delayed images (>24 h) obtained using ^{67}Ga citrate.
13. (a) ^{18}FDG uptake in tumors is due to their high glycolytic rates. Poorly differentiated tumors have greater uptake than well-differentiated tumors; this is the reason that the standard uptake value (SUV) has prognostic value.

14. (d) ^{18}F FDG is a glucose analog that is taken up preferentially in many tumor types due to their higher glycolytic rates. Patients should fast in preparation for the exam because hyperglycemia can reduce tumor uptake of ^{18}F FDG. ^{18}F FDG-PET/CT is routinely used for lung cancer staging, treatment planning, and treatment response evaluation. However, the majority of prostate cancer has low FDG uptake.
15. (b) If GI bleeding study and ^{18}F FDG-PET/CT are performed on the same day, ^{18}F FDG-PET/CT should be performed first as residual ^{99m}Tc activity will not interfere with PET imaging.
16. (a) ^{18}F FDG-PET/CT should not be performed first as any residual ^{18}F FDG activity will interfere with gastric emptying study imaging. If gastric emptying study is performed first, the patient will not meet NPO requirement for ^{18}F FDG-PET/CT.
17. (b) Malignant cells have altered glucose metabolism; hence, ^{18}FDG can be used for tumor imaging.
18. (b) Hodgkin's disease is a kind of lymphoma; it is most common in young adults. One of the first complaints is painless, enlarged lymph nodes. It has been imaged using ^{67}Ga citrate and more recently ^{18}F FDG.
19. (d) F-18 fluciclovine is an FDA-approved imaging agent for prostate cancer. More recently, FDA approved Pylarify® (F-18 piflufolastat) for PET imaging in men with prostate cancer.
20. (a) Sixty-eight minutes is the half-life of gallium-68 DOTATATE.
21. (d) The recommended dose is weight based with a maximum dose of 5.4 mCi.
22. (a) Gallium-68 is a generator-produced PET imaging agent.
23. (b) Gallium-68 DOTATATE binds to somatostatin receptors in neuroendocrine tumors.
24. (a) The recommended dose is 10 mCi.
25. (b) FDG is not recommended to evaluate prostate cancer in most cases.
26. (c) PSMA is expressed in normal prostate tissue. All others are correct.

27. (a) Axumin PET images are obtained in 3–5 min post-tracer injection. All other images are obtained approximately 60 min after tracer injection.
28. (d) 18F-fluoroestradiol PET/CT is mainly used for the evaluation of heterogeneity of estrogen receptor (ER) expression in breast cancer.

Appendix 13: Answers to Chapter 13

31

1. (a) ^{67}Ga emits gamma rays of 93, 184, 269, and 394 keV. ^{111}In emits gamma rays of 172 and 247 keV, so choice (c) is not true. Labeled leukocytes are used for infection imaging rather than neoplasms. Medium-energy collimation is used for both.
2. (b) It is best to use a radiopharmaceutical with less gastrointestinal excretion; therefore, ^{67}Ga should be excluded. Octreotide is used for tumors with somatostatin receptors. ^{99m}Tc-labeled leukocyte imaging is used for detecting infection but with undesirable abdominal activity. ^{111}In-labeled leukocytes would be useful with very low GI activity.
3. (e) Leukocytes are labeled with ^{111}In and ^{99m}Tc for infection imaging. ^{67}Ga-labeled leukocytes are not used clinically.
4. (b) It can be assumed, because of the high activity present at 24 h, that the image was not obtained using a ^{111}In radiopharmaceutical. ^{99m}Tc-labeled WBC imaging is typically obtained the same day within a few hours and with higher abdominal activity. One would expect to see bowel and nasopharyngeal and mediastinal activity if ^{67}Ga were used. Soft tissue uptake is not expected if ^{99m}Tc sulfur colloid is used.
5. (b) Inadvertently labeled platelets may show thrombosis.
6. (c) To prevent clotting, either heparin, ACD solution, or other anticoagulant should be used in the collecting syringe. Heparin is less desirable as some patients display hypersensi-

E. Mantel et al., *Nuclear Medicine Technology*, https://doi.org/10.1007/978-3-031-26720-8_31

tivity. Small-bore needles may damage cells. No syringe shield is needed.

7. (b) Patients who are leukopenic have too few white blood cells, so labeling their leukocytes may not be feasible, ruling out choices (a) and (c) (although donor cells could possibly be used). ^{99m}Tc sulfur colloid is used for bone marrow imaging but not for infection screening. ^{67}Ga citrate or FDG-PET/CT is an alternative choice.
8. (b) Gravity sedimentation is the settling of solids to the bottom of a liquid based on weight and does not require centrifugation or magnets. It may be used for separation of erythrocytes and is often facilitated by the addition of hetastarch to increase the attraction of erythrocytes to one another.
9. (a) Five hundred μCi to 1.0 mCi of ^{111}In oxine-labeled leukocytes is a common dose range for imaging infection. Using ^{99m}Tc exametazime allows more activity to be administered and offers a shorter injection to imaging time.
10. (c) The immediate and blood pool phases of bone scans can be used to differentiate cellulitis and osteomyelitis; ^{67}Ga citrate is a helpful addition in cases of complicated osteomyelitis.
11. (a) If scanning for infection, the ^{111}In leukocyte exam could be used, but the radiation dose to the patient might be avoided by obtaining further images over time to confirm whether the activity is changing as gallium is excreted through bowel. Hydration and voiding are not helpful in ruling out bowel excretion of gallium, neither are subtracted liver/spleen images. Laxatives or enemas may also be of help.
12. (d) ^{67}Ga provides high sensitivity for the diagnosis of vertebral osteomyelitis.
13. (b) ^{111}In-WBC scan is the best choice. ^{99m}Tc-WBC scan and ^{67}Ga citrate scan both have physiologic activity in the abdomen and thus complicate the study. Bone scan has no role here.
14. (d) ^{99m}Tc-WBC scan is preferred in pediatric patients, due to lower radiation.

15. (b) ^{99m}Tc sulfur colloid imaging helps to determine if increased ^{111}In-WBC activity is due to excess of bone marrow tissue, and thus increase the specificity of ^{111}In-WBC scan.
16. (e) ^{99m}Tc-WBC has a shorter half-life, higher dose can be used, and imaging starts 2–4 h after tracer injection. However, ^{111}In-WBC scan and ^{99m}Tc-WBC scan share similar mechanism in detecting infection.
17. (b) Spleen has the highest physiologic activity in ^{111}In-WBC scan (also in ^{99m}Tc-WBC scan).
18. (e) ^{99m}Tc sulfur colloid is cleared by mononuclear phagocyte system, including the liver, spleen, and bone marrow.
19. (e) Acute osteomyelitis will show positive findings on all three phases of bone scan.
20. (d) A negative three-phase bone scan essentially rules out the diagnosis of osteomyelitis, and thus is very helpful in the differential diagnosis. In contrast, a positive three-phase bone scan may be caused by multiple etiologies.
21. (d) After hip replacement, MDP uptake in most cases returns to normal in 12 months, but it is not rare to see persistent uptake lasting for more than 2 years. However, a negative bone scan essentially rules out the diagnosis of loosening/infection.
22. (e) FDG-PET/CT is able to evaluate infection and tumor at the same time and is the preferred study for patients with fever of unknown origin. ^{67}Ga citrate scan can also be used.
23. (e) All of the above.
24. (d) While noncaloric beverages are allowed, NPO for at least 4 h is also required.
25. (b) Areas of physiologic uptake of ^{67}Ga citrate include the nasopharynx, lacrimal and salivary glands, liver, and pediatric epiphyses.
26. (e) Low-energy collimation is generally used for photopeaks <150 keV. Using insufficient collimation will result in increased septal penetration with the concomitant negative effect on resolution. The decreased septal depth will increase sensitivity.

Appendix 14: Answers to Chapter 14

32

1. (c) The percent uptake is calculated by the following formula:
   Thyroid counts − thigh counts/(capsule counts − background) (decay factor) × 100%. Because a standard is not being used, the capsule has been counted before administration, background is subtracted, and activity is multiplied by the decay factor. If a standard were used, the formula would be:
   Thyroid counts − thigh counts/counts in standard × 100%.
2. (a) The normal range for 6-h uptake is about 6–18%; for 24 h, it is 20–30%, although these numbers vary according to the laboratory.
3. (d) Because of the proximity to the bladder, counts will be higher, and this will result in a falsely low uptake.
4. (d) Kelp tablets contain iodine and therefore may affect measured values of thyroid uptake.
5. (d) Choices (a), (b), and (c) may affect measured values of thyroid uptake, but beta-blockers should not.
6. (b) Thyrotropin is also known as thyroid-stimulating hormone (TSH). It is secreted by the anterior pituitary and causes the thyroid to release thyroxine (T4).
7. (b) TRH is a thyrotropin-releasing hormone which is secreted by the hypothalamus and stimulates the secretion of TSH by the anterior pituitary. The thyroid releases the hormones T3 (triiodothyronine) and T4.

E. Mantel et al., *Nuclear Medicine Technology*,
https://doi.org/10.1007/978-3-031-26720-8_32

8. (a) Most people have four, but a small percentage of people may have more or less.
9. (d) The major salivary glands include the parotid, sublingual, and submaxillary glands. There are also minor glands in the cheeks, lips, mouth, and throat.
10. (e) Sometimes, one of these imaging agents is paired with pertechnetate imaging to create images that will allow subtraction of the thyroid.
11. (d) The pinhole may be used for imaging, but the flat field is used for uptake.
12. (a) Lugol's solution is iodine and should not be administered. Voiding is not necessary before administration of the radiopharmaceutical, nor is withholding caffeine. Most laboratories have a policy of fasting before and for a few hours following the dosing, to enhance digestion of the dose.
13. (e) See explanation for Question 1, this chapter.
14. (b) Cytomel is a synthetic form of T3 and will affect iodine uptake, so if images are needed, they can be obtained using thallium. Any uptake values obtained would be falsely low; this medication should be stopped at least 2 weeks before uptake and scanning using radioiodine.
15. (b) Pertechnetate localizes in the thyroid by active transport; it is trapped in the gland, but, unlike iodine, not organified.
16. (d) Imaging may be performed in all cases, but images taken at 6 h after radioiodine ingestion will have a higher body background relative to 24-h images. Those obtained at 24 h will have a lower count rate, but images may be superior due to the decreased body background.
17. (d) The xiphoid process will not likely be in the FOV.
18. (c) Activity that localizes in the salivary glands and is secreted may be swallowed resulting in esophageal activity. If this is seen on a pertechnetate scan, the artifact can be distinguished from the pyramidal lobe by having the patient drink.
19. (b) Whole-body scanning with ^{131}I to rule out metastatic disease is often performed after thyroidectomy. The dose used is 3–5 mCi. Thyroid uptake and scanning are typically performed using ^{123}I.
20. (b) See Question 10, this chapter.

21. (c) Pertechnetate is used for imaging the salivary glands, often to evaluate function or rule out masses.
22. (b) ^{99m}Tc sestamibi washes out of the thyroid over time, and parathyroid adenomas, which may appear more intense than the thyroid on early images, often retain activity on delayed images.
23. (a) The majority of hot nodules seen on thyroid scans are benign.
24. (a) Bradycardia means a slowing of the heart rate; rapid heart rate may be a symptom of hyperthyroidism. Cold intolerance is one of the symptoms of hypothyroidism. Exophthalmos, or protrusion of the eyeballs, may be seen in hyperthyroid patients.
25. (c) The therapeutic dose of radioiodine for thyroid cancer is 150–200 mCi and may be higher if distant metastases are being targeted.
26. (b) See explanation for Question 2.
27. (c) See explanation for Question 28, this chapter.
28. (c) The thyroid appears as a butterfly shape on imaging, with the isthmus often not visualized, or displaying decreased activity relative to the right and left lobes. A slightly larger right lobe is commonly seen and is usually a normal finding.
29. (a) Ingested iodine is absorbed in the small intestine and transported to the thyroid where it is needed for the synthesis of T3 and T4.
30. (a) The thyroid secretes much more thyroxin (T4) than triiodothyronine (T3). Both are secreted by the follicular cells of the gland. If too little T4 is secreted, myxedema will result.
31. (d) Hyperparathyroidism may be caused by hyperplasia of the glands or by a tumor and will result in increased PTH secretion. This causes an increased risk of bone fractures because PTH, when elevated, will foster the removal of calcium from bone. The high PTH can also affect the nervous system.
32. (b) In cases of extremely rapid iodine turnover, only the 6-h uptake may be abnormal, for example, in Graves' disease.
33. (a) The SNM recommends that patients wait 2–4 weeks after examinations using iodinated contrast agents before having a thyroid uptake measurement, but many departments recom-

mend up to 6 weeks. If the myelogram was performed using Pantopaque, 2 years' delay may be needed, but this is not often used because of the possible association with arachnoiditis.

34. (b) Lemon juice stimulates the salivary glands to secrete saliva and pertechnetate, with a resulting decrease in activity in the glands.
35. (e) Ectopic thyroid tissue may be present in many places; in this case, ^{131}I imaging may be helpful. It is not taken up by the salivary glands, the long half-life allows imaging after the target-to-background ratio has become higher, and there has been urinary excretion of much of the dose.
36. (d) It is typical to see salivary glands superior to the thyroid on images obtained using pertechnetate. If they are not seen, it may be because the patient has Graves' disease or because salivary gland function is compromised.
37. (d) The LEAP collimator is ideal for imaging with technetium if magnification of the gland is not needed. Images obtained with the pinhole collimator may follow. Uptake counting requires a flat-field collimator, and medium or high energy may be used for ^{131}I.
38. (a) Both radioiodine and pertechnetate will cross the placenta.
39. (b) See explanation for Question 1, this chapter.
40. (a) The calculation of dose for radioiodine therapy takes the uptake into consideration.
41. (d) Amiodarone is an antiarrhythmic drug and contains high level of iodine.
42. (d) The findings are likely real. The technologist should consult a physician at this time.
43. (d) All others are wrong.
44. (a) The typical dose is 20 mCi ^{99m}Tc sestamibi.
45. (c) Parathyroid doses are administered via IV bolus.
46. (b) Both ^{99m}Tc sestamibi and ^{123}I are administered for dual-isotope parathyroid imaging.
47. (c) 5 mCi $^{99m}TcO_4^-$ is administered for a thyroid scan.
48. (a) 100 uCi ^{131}I is administered for a diagnostic thyroid scan.
49. (c) ^{131}I decays via beta decay.
50. (b) The energy of ^{131}I is 364 keV.

51. (a) Whole-body ^{131}I imaging is performed using 2–5 mCi ^{131}I.
52. (b) The half-life of ^{123}I is 13 h.
53. (b) Imaging should be initiated within 30 min of the administration of $^{99m}TcO_4^-$.
54. (d) All of the above are correct, thyroid function, functional status of nodules, and size and location of thyroid tissue.
55. (b) $^{99m}TcO_4^-$ is administered intravenously.
56. (c) ^{123}I is administered orally.

Appendix 15: Answers to Chapter 15

33

1. (d) ^{131}I is used to treat thyroid cancer and hyperthyroidism by ablation. Graves' disease is the most common cause of hyperthyroidism.
2. (d) The typical adult dose is 500 mCi.
3. (b) Somatostatin analogs need to be discontinued prior to the therapy administration. Long-acting somatostatin analogs need to be discontinued for 4 weeks, and short-acting somatostatin analogs need to be discontinued for at least 24 h.
4. (b) Y-90 microspheres are administered via intra-arterial injection.
5. (b) In vivo is the only RBC tagging method that does not require blood handling.
6. (c) ^{32}P chromic phosphate is useful in treating malignant ascites. ^{32}P sodium phosphate is used to treat polycythemia vera. ^{89}Sr (as well as ^{32}P sodium phosphate and ^{153}Sm-EDTMP) is used to treat malignant bone pain by localizing where there is bone mineral turnover. ^{131}I may be used to treat Graves' disease.
7. (c) ^{89}Sr chloride can be administered by direct venous access, but as it is a beta emitter, it is a better option to access the vein with an intravenous line and check the patency of the system before injection. If the dose is infiltrated into the tissue surrounding the vein, the tissue will be unnecessarily irradiated.

E. Mantel et al., *Nuclear Medicine Technology*,
https://doi.org/10.1007/978-3-031-26720-8_33

8. (b) ^{89}Sr and ^{32}P are beta emitters and will be effectively shielded with the plastic syringe. Using a lead syringe shield will cause bremsstrahlung.
9. (b) The half-life of radium-223 is 11.4 days.
10. (a) The tagging efficiency using the in vivo tagging method is 85%.
11. (c) 95.3% is emitted as alpha rays.
12. (a) True, patients with a lung shunt of greater than 20% is a contraindication for the therapy administration.
13. (a) TheraSphere® microspheres are glass.
14. (c) In vitro labeling utilizes a centrifuge to separate the blood components. Neither in vivo nor modified in vivo utilizes a centrifuge.
15. (a) The typical dose regimen is four 200 mCi doses every 8 weeks. Dose may be adjusted based on the patient's blood cell counts. Patient's counts must be evaluated after every treatment. Myelosuppression can occur, and if it does, doses may be modified or discontinued.
16. (a) Mapping is used to determine the percent of shunting from the liver to the lungs. See answer to Question 12 for additional information.
17. (a) A beta particle with energy of 2 MeV has a range of about 1 cm in soft tissue. That means it will expand its energy locally, regardless of tissue type. Some beta emitters are not useful for imaging because of their short range, although higher energy beta emitters may be imaged, for example, ^{131}I. Half-life varies according to the isotope; ^{32}P has a half-life of 14.3 days, but ^{60}Co has a half-life of 5.26 years.
18. (c) Polycythemia means too many red blood cells. Polycythemia vera is a disease of the bone marrow, which may be treated using ^{32}P sodium phosphate.
19. (d) Red cell volume and hematocrit are measures of red blood cells in whole blood. Hemoglobin is a specific iron-rich protein that carries oxygen, so it does not directly measure the number of red blood cells. Hemoglobin may be decreased with or without a decrease in the number of red blood cells.

20. (a) Red blood cells, or erythrocytes, are produced in bone marrow, live for about 120 days, and are then phagocytized, mainly by the spleen. This is why damaged red blood cells may be used to image the spleen.
21. (b) Blood contains red blood cells, white blood cells, and platelets (thrombocytes). Red blood cells are erythrocytes, leukocytes are white blood cells, and granulocytes are a kind of white blood cells.
22. (a) In the absence of an anticoagulant, the blood will separate into serum and a clot containing cells and coagulation proteins.
23. (d) B cells initiate the production of antibodies, T helper cells mediate cellular immunity, and macrophages (a kind of white blood cell) phagocytize pathogens.
24. (b) Plasma contains mostly water, with dissolved salts and proteins. In many patients, water will move from the bloodstream into the extravascular space in the legs as a result of standing, thus changing the plasma volume.
25. (b) Since the limit for exposure to nonoccupational individuals is 2 mrem/h, we can use the formula below:

$$\text{Total dose} = \text{Dose rate} \times \text{Time}$$

$$\text{Rearrange to: Dose rate} \times \text{Time} = \text{Total dose}$$

$$10\,\text{mrem}/\text{h} \times \text{Time} = 2\ \text{mrem}$$

$$\text{Time} = \text{mrem}/\left(10\,\text{mrem}/\text{h}\right)$$

$$0.2\ \text{h}$$

26. (b) The formula below can be used:

$$\text{Initial dose rate}/\text{initial mCi} = \text{New dose rate}/\text{mCi remaining}$$

$$\left(60\ \text{mrem}/\text{h}\right)/45\ \text{mCi} = \left(30\ \text{mrem}/\text{h}\right)/\text{mCi remaining}$$

$$\text{Remaining} = \left(\left(30\ \text{mrem}/\text{h}\right)\times 45\ \text{mCi}\right)/60\ \text{mrem}/\text{h}$$

$$\text{mCi remaining} = 22.5\,\text{mCi}$$

Of course, these numbers are meant as an exercise and do not reflect actual practice since the half-life of ^{131}I is 8.06 days.

27. (d) The exposure limit for nonoccupational individuals is 2 mrem/h. The dose rate in the question is 5 mrem/h, so if he or she stays for one-half hour, he or she receives a dose of 2.5 mrem. If he or she doubles her distance from the patient, he or she will decrease his or her dose by one-fourth of the original, so he or she would receive 1.25. If he or she stays for only 20 min, he or she receives a dose of 1.7 mrem ((5 mrem/1 h) × 0.3 h = 1.5), and if he or she returns the next day, he or she will receive much less because of the physical half-life. Therefore, any of these options are effective for reducing the dose to meet NRC requirements. However, principles of ALARA state that the dose should be as low as is reasonably achievable. So, he or she should not interview the patient at all or wait as long as possible before performing the interview.
28. (e) 10 CFR 35.75 states that a patient can be released if the total effective dose equivalent to any other individual from exposure to the released individual is not likely to exceed 5 mSv (0.5 rem). Patients who have received (or determined to have retained) <33 mCi of ^{131}I may be released.
29. (d) Although bathroom surfaces are most likely to be contaminated, anything the treated patient comes in contact with has the potential to become contaminated.
30. (d) All patients who are scheduled for nuclear medicine examinations or therapy should be screened for pregnancy and breastfeeding. Radioiodine will be better absorbed if the patient has been fasting. Patients should avoid iodine-containing substances (including contrast agents and food), thyroid hormones, and medications that would interfere with iodine uptake into the thyroid. The necessary delay before imaging after stopping various medications varies, and the SNM Guidelines provide detailed information about this.
31. (d) The dose necessary for ^{131}I therapy for hyperthyroidism takes into consideration the weight of the gland (normally 15–20 g, but this is estimated by palpation and imaging) and the uptake. In general, 80–200 μCi of ^{131}I per gram of thyroid tissue is desirable. One formula used to calculate dosage is thyroid weight/24-h uptake × 10 = dose of ^{131}I in mCi. Other methods consider nodularity as well.

32. (c) The uptake is multiplied by the dose administered to find the amount of activity in the gland; this is then divided by the weight of the gland to find the activity/gram:

$$20\ \mu\text{Ci} \times 0.25 = 5\ \mu\text{Ci}$$
$$5\ \mu\text{Ci} / 45\ \text{g} = 0.1\ \mu\text{Ci} / \text{g}$$

33. (d) ^{131}I is a pure beta emitter; the beta emission is responsible for the local effect of treatment. Radioiodine therapy depends on functioning thyroid tissue, so choice (a) is incorrect. The half-life of ^{131}I is 8.06 days, which is desirable; a beta emitter with a very short half-life would not have the same treatment benefit.
34. (e) Hyperthyroidism may be treated with surgery, radioiodine, or drugs to disrupt the production of thyroid hormones. Thyroid storm is potentially a life-threatening condition requiring emergency treatment caused by excessive thyroid hormone; symptoms include high fever, increased heart rate, and sweating.
35. (b) During the wash method of red cell labeling, the free chromate ion is removed and so is the possibility that they will label circulating red blood cells after reinjection of the labeled preparation. This is a major advantage to the method. Anticoagulant is still used in the initial blood draw, and the same amount of radioactive chromium is used as in the ascorbic acid method. It also negates the need for adding ascorbic acid to reduce the free chromate ion.
36. (b) ^{32}P chromic phosphate is a colloid and therefore will be phagocytized by the Kupffer cells of the liver if injected intravenously. Because it is a pure beta emitter, it will cause local radiation damage.
37. (c) Patients need to continue efforts to decrease radiation exposure to others and themselves, including limiting contact with others, frequent voiding, and flushing multiple times after using the toilet. Patients do not need to collect and store excreta.
38. (a) The formula used in the explanation to Question 26 is applied to find 27 mCi remaining, so the patient can be released. See explanation for Question 28 regarding the release of patients.

39. (b) The patient must have a private room with a private bathroom or may share a room and a bathroom with another patient who also cannot be released under 10 CFR § 37.75 (see also 10 CFR 35.315).
40. (c) The NRC requires that these records be kept for 3 years.
41. (b) Because the injected dose volume must be determined by weighing the syringe before and after injection, one should perform the injection without rinsing the syringe with blood.
42. (a) Because the patient dose is 40–60 μCi/kg, the dose is usually not high enough to require hospitalization.
43. (c) ^{90}Y ibritumomab tiuxetan is a monoclonal antibody used in the treatment of non-Hodgkin's lymphoma. ^{82}Rb decays by positron emission and also electron capture and is used in PET imaging. ^{32}P chromic phosphate is a beta emitter used for intracavitary therapy. ^{153}Sm microspheres are being developed for radiosynovectomy but are not FDA approved at the time of writing.
44. (d) ^{90}Y ibritumomab tiuxetan and ^{131}I tositumomab are labeled monoclonal antibodies produced using mouse cells and will likely cause an allergic response in those patients who have previously demonstrated allergy to mouse proteins.
45. (d) When ^{99m}Tc pertechnetate is used for thyroid scintigraphy, the dose is higher and the radiation exposure is lower, the imaging is obtained 10 min after injection, and no special patient preparation is needed. However, no thyroid uptake ratio can be calculated.
46. (d) See the above answer.
47. (d) Breastfeeding should be stopped and cannot be resumed if ^{131}I is used. If ^{123}I is used, breastfeeding can be resumed 1 week later.
48. (a) Half-life is 11.4 days.
49. (c) Metastatic prostate cancer
50. (a) Intravenous injection
51. (a) The recommended dose is 1.49 μCi/kg of body weight.
52. (b) Pheochromocytoma is an indication for Azedra® therapy.
53. (b) Heparin is an anticoagulant.
54. (b) The half-life is 8.1 days.
55. (b) Azedra® is administered intravenously via infusion pump.

56. (c) An anticoagulant is not used when labeling RBCs with the in vivo method.
57. (a) The half-life of Lu-177 is 6.6 days.
58. (a) The typical dose regimen is six injections at 4-week intervals.
59. (c) Thyroid-blocking agents such as SSKI or Lugol's solution need to be administered before the Azedra® therapy administration.
60. (b) Lutathera is administered via intravenous infusion over 30–40 min.

34 Appendix 16: Answers to Chapter 16

1. (d) While a technologist has to take care of a patient, he or she is responsible to be alert for and take steps to avoid problems with IV lines, catheter bags, etc.
2. (b) Ambulatory means able to walk.
3. (c) Protective barriers are a part of universal precautions set out by the Centers for Disease Control and Prevention and include gowns, gloves, and masks as necessary. There is a minute risk of spread of HIV via contact with tears, feces, nasal secretions, sputum, urine, and vomit, unless there is visible blood in them. However, in some cases, it may be necessary for technologists to use gowns or masks to protect their mucous membranes and skin if there will be exposure to other body fluids or those listed above when they contain blood. Obtaining a sexual history may be relevant for the physician but does not help a technologist control disease spread.
4. (b) Bleeding following an IV injection is halted by applying pressure. In cases of severe or prolonged bleeding, a physician should be alerted.
5. (b) NPO stands for *nil* per os, which means nothing by mouth.
6. (d) Patient restraint devices should not restrict circulation, and patient comfort is important. Buckles on straps, etc. may produce attenuation artifacts.
7. (d) Aphasia refers to difficulty with or complete inability to use or comprehend words.

E. Mantel et al., *Nuclear Medicine Technology*,
https://doi.org/10.1007/978-3-031-26720-8_34

8. (c) Anaphylactic reaction is a systemic allergic reaction caused by a second exposure to an antigen, so this is unlikely to occur in multiple patients. Another type of allergic reaction (which may be mild) would also not likely be present in multiple patients. The dose necessary to cause radiation sickness would be much higher than that likely to be contained in the vial. A pyrogenic reaction is one causing fever and could be the result of pyrogens (microbial or nonmicrobial substances that produce increases in body temperature) in the prepared radiopharmaceutical and therefore could affect all patients receiving doses withdrawn from that vial.
9. (b) A gastrointestinal bleeding study does not require that the patient be NPO prior to exam, but this is part of the preparation for all of the other choices.
10. (d) The Centers for Disease Control and Prevention recommends the healthcare professional treating patients for whom droplet and airborne precautions are needed use masks when treating the patients, put such patients in private rooms, and that the patient wear a mask when being transported or treated outside his or her room.
11. (b) Standard precautions (now known as universal precautions) apply to all patients regardless of their infection status and include the use of gloves, handwashing, barrier protection (depending on the anticipated exposure), and safe injection procedure.
12. (c) Nosocomial refers to being acquired in a hospital.
13. (d) Patient identity should always be ascertained before performing any patient procedure. Identification bands, charts, and verbal communication should all agree with the identity of the patient before proceeding.
14. (d) Regardless of the seizure type, patients should not be restricted and nothing should be inserted into their mouth as this will increase the risk of injury. Removing objects that could potentially injure the patient and prevention of injury are priorities for the healthcare provider during the seizure. It is not necessary to start CPR in cases of seizure (unless, of course, the seizure is provoked by a stroke or other serious

medical conditions and the patient's heartbeat and breathing cease).

15. (d) Infectious waste should be disposed of using the universal biohazard symbol, which is recognized by the World Health Organization and numerous national agencies. Sharps should be disposed of using puncture-resistant containers, and all containers and bags should be leakproof.
16. (d) While nasal secretions do not require universal precautions, vaginal secretions do, and the vaginal lining is a mucous membrane. Hence, mucous membranes are included in the correct answers. In practice, nuclear medicine technologists rarely have contact with mucous membranes.
17. (e) The Centers for Disease Control and Prevention recommends changing gloves between patients and between procedures on a single patient if moving from a contaminated area to a clean area. There is no specific frequency recommendation.
18. (d) The Centers for Disease Control and Prevention recommends handwashing or another method of hand hygiene, before and after direct contact with a patient and after gloves are removed, in addition to other circumstances.
19. (b) Single-use sharps should not be recapped before disposal as the proper disposal container is puncture resistant and recapping will increase the risk of puncturing the skin of the user.
20. (b) The International Commission on Radiological Protection recommends delaying breastfeeding for 12 h after ^{99m}Tc-labeled radiopharmaceuticals except for labeled red blood cells, phosphonates, and DTPA for which 4 h of interruption is recommended. Three weeks is recommended after ^{201}Tl-, ^{67}Ga-, and ^{125}I-labeled compounds. It is recommended that breastfeeding be completely discontinued after ^{131}I therapy.
21. (b) While some investigational radiopharmaceutical use will require that the patient is hospitalized and some could theoretically require storage of urine until decay, the only statement that is true regarding all uses of investigational drugs is that patients must give informed consent.

22. (d) The use of radiopharmaceuticals is not encouraged during pregnancy. Iodine, and therefore radioiodine, crosses the placenta from the mother to the child, and therefore, radioiodine therapy can damage the baby's thyroid. Women are advised to delay pregnancy for a year following radioiodine therapy. The International Atomic Energy Agency recommends lower doses followed by hydration and frequent voiding to reduce fetal exposure for pregnant women who have undergone examinations using radiopharmaceuticals. Hydration without frequent voiding will increase the dose to the fetus. The total dose equivalent allowed to the fetus from occupational exposure of the mother is 0.5 rem.

Appendix 17: Answers to Chapter 17

35

1. (d) ^{14}C decays by beta emission. All of the other choices listed decay by positron emission and some also by electron capture.
2. (d) The atomic mass number is the number of particles in the nucleus. Because the proton is "converted" to a neutron, the number of protons decreases by one, and the number of neutrons is increased by one, so the net effect is no change to the atomic mass number. The atomic number, which is the number of protons in the nucleus, decreases by one.
3. (a) ^{82}Rb has a half-life of 1.3 min, ^{13}N has a half-life of 10 min, ^{15}O has a half-life of 122.2 s, and ^{18}F has a half-life of 109 min.
4. (e) Increased physical activity in the days prior to scanning will increase muscle uptake of ^{18}F-FDG. Catheterization can decrease the activity in the pelvis.

   Increased glucose level can decrease tumor uptake, so many institutions reschedule a patient's PET scan if their blood glucose level exceeds 200 mg/dL.
5. (b) ^{18}F-FDG will accumulate in the brain, urinary tract, and myocardium. Bone activity can be seen with ^{18}F-NaF.
6. (b) Many positron emitters are cyclotron produced but not all. ^{82}Rb is eluted from generators containing ^{82}Sr.
7. (c) ^{18}F-FDG-PET is used for the indications in choices (a), (b), and (d). Detecting *H. pylori* is accomplished using a

E. Mantel et al., *Nuclear Medicine Technology*,
https://doi.org/10.1007/978-3-031-26720-8_35

radioactive carbon breath test. (See Chap. 10 ("Gastrointestinal Tract Scintigraphy"), Question 38.)

8. (d) The normal distribution of ^{18}F-FDG in the brain is related to both blood flow and metabolic rate, with gray matter showing the greatest uptake. It is a glucose analog, has a half-life of 109 min, and is taken up by normal myocardium.
9. (c) Because FDG is a glucose analog, this study shows a map of glucose distribution.
10. (b) With a physical half-life of 109 min and a biological half-life of 6 h, the effective half-life for ^{18}F is approximately 1.4 h. Nonetheless, many imaging departments recommend that nursing mothers refrain from nursing for 24 h following a PET study with this isotope, although this varies greatly among labs. Some laboratories recommend as little as 6-h interruption after imaging, so it is completely eliminated from the body in 14 h.
11. (b) Despite the higher energy of annihilation photons relative to other photons detected in nuclear medicine imaging, attenuation correction is even more important in PET because it requires detection of both annihilation photons from a single decay. Because these photons are oppositely directed, the pair of annihilation photons must pass through the entire width of the patient to reach the detector ring. Attenuation correction can be accomplished using a transmission scan from either a sealed source or a CT, which measures attenuation through all lines of response.
12. (d) A blank scan is performed using a transmission source in an empty field of view. This is used to monitor system stability and is also needed along with the transmission scan to perform attenuation correction. In CT systems, a blank scan is often referred to as an air cal.
13. (c) The simultaneous detection of annihilation photons originating from a single positron is called a true coincidence. The simultaneous detection of annihilation photons after one (or both) of the photons has undergone Compton scatter is called a scatter coincidence. Because Compton-scattered photons change direction, this type of coincidence will be assigned to an incorrect line of response (LOR). The resulting misposi-

tioned events decrease image contrast. The simultaneous detection of photons originating from different positrons is called a random coincidence, and the random rate increases as the square of the amount of activity present in the FOV. This effect must be corrected for in order to obtain quantitative data.

14. (c) Scintillation crystals that have been used in PET cameras include NaI(Tl), LSO, BGO, and GSO. Lead sulfate, $PbSO_4$, is the white powder often seen on the electrodes of car batteries.
15. (b) Reconstruction of PET images can be performed using filtered back projection or iterative reconstruction schemes. Coincidence detection refers to the process by which photons are accepted or rejected as originating from a single annihilation event. K-space filling is a part of MRI reconstruction. Block detection refers to the division of scintillators into separate channels using a material that will not transmit light.
16. (b) The deflected photon can be detected in a position that changes the assignment of the LOR. Choice (a) is not correct because an unpaired photon will not contribute to the image. Lines of response are lines and do not contain angles. In Compton scatter, the deflected photon transfers (loses) some of its energy to an electron. Therefore, a deflected photon's energy is always lower than that of an undeflected photon.
17. (b) ^{13}N ammonia can be used to image myocardial perfusion, and ^{18}F-FDG can be used to assess myocardial metabolism. In ischemia, the myocardial metabolism of glucose increases, relative to fatty acids, so one would expect to see increased uptake of ^{18}FDG in an ischemic area.
18. (a) Because there will be less attenuation, more true coincidences will be detected.
19. (b) Any activity that utilizes a muscle has the potential to cause an increase in the uptake of FDG in that muscle.
20. (c) Areas of inflammation often show increased activity on ^{18}F-FDG images.
21. (b) Dietary instructions prior to imaging vary greatly, but the minimum necessary fast is agreed by most to be 4 h prior to

injection. Avoiding intense physical activity is also recommended for the days prior to imaging.

22. (d) ^{18}F-FDG is taken up by tumors according to their glycolytic rate, which is often higher than that of normal tissues, and once localized in tumor cells, it is phosphorylated and so remains there.
23. (c) The dose recommended by the Society of Nuclear Medicine Procedure Guideline for whole-body FDG-PET scanning is 10–20 mCi injected intravenously (370–740 MBq).
24. (a) See explanation for Question 4.
25. (a) Isobars are atoms of different elements; they have different numbers of protons but have the same mass number (number of protons plus neutrons). Isotopes are members of the same element, so they have the same number of protons but have differing numbers of neutrons. Isotones have the same number of neutrons.
26. (c) The positron range is the distance that the positron travels before undergoing annihilation with an electron. The distance is radionuclide dependent and is, on average, 0.6 mm in water for ^{18}F and 5.9 mm in water for ^{82}Rb. Because PET scanners detect the products of positron annihilation rather than the positron itself, the line of response will be assigned a slight distance away from the actual site of positron emission.
27. (a) Regardless of how a patient receives his or her nutrition, increased glucose level may decrease tumor uptake because glucose will compete with FDG for cellular uptake. Of course, the decision to remove nutrition is up to the clinician.
28. (b) A random coincidence occurs when two annihilation photons from different annihilations are detected within the coincidence timing window and treated as a true coincidence. A true coincidence refers to the detection of two annihilation photons from a single annihilation event. A scatter coincidence is the term for detection of two annihilation photons from a single annihilation event, where one or both photons have undergone Compton scatter, resulting in a change in direction.

29. (a) The angle difference from 180° is typically ±0.25% and negatively affects spatial resolution. The effect is increased with increasing diameter of the scanner.
30. (d) Muscle uptake may be the result of muscle contraction during the uptake of ^{18}F-FDG, so premedicating the patient with diazepam (or other anxiolytic drugs), keeping the patient supine and relaxed, and minimizing the use of facial and neck muscles can decrease this uptake. Muscular uptake in the neck can be difficult to differentiate from lymph nodes.
31. (b) Electronic collimation for PET means that pairs of photons striking the detector ring within a short timing window are assumed to come from the same positron decay, so the event is positioned by drawing a straight line (line of response) between the two detectors.
32. (c) Standardized uptake value (SUV) is the measured activity concentration divided by the average activity concentration in the patient's body, commonly calculated as the administered activity divided by body weight. The measured activity concentration can be obtained from a single voxel, as in the case of maximum SUV (SUV_{max}), or from a volume of interest (VOI), as in the case of average SUV or peak SUV. Comparing SUVs in pre- and posttreatment images may show response to therapy.
33. (b) If any of the dose is extravasated, it will be cleared by the lymph nodes. This may cause axillary uptake to complicate scan interpretation. Hence, the injection site should be contralateral to the area in question or in the foot.
34. (a) The recommended dose is 0.054 mCi/kg of body weight, with a maximum dose of 5.4 mCi.
35. (c) F-18 PSMA is used to image patients with prostate cancer.
36. (a) Netspot® imaging is used for localization of somatostatin receptor-positive neuroendocrine tumors.
37. (b) Prostate cancer is imaged with Axumin (fluciclovine F-18).
38. (a) F-18 PSMA is administered via intravenous injection.
39. (a) Imaging with fluciclovine F-18 should begin 3–5 min after administration.

40. (a) True, both adult and pediatric patients can be imaged with Netspot®.
41. (b) 9 mCi is the typical dose administered for F-18 PSMA PET imaging.
42. (c) The delay between injection and scanning with F-18 PSMA is 60 min.
43. (a) True, Ga-68 DOTATATE is generator produced.
44. (a) The typical dose of fluciclovine F-18 for PET imaging is 10 mCi.
45. (a) True, there are no contraindications to using Axumin.
46. (b) False, patients do not have to be NPO prior to F-18 PSMA PET imaging.
47. (c) The delay between injection and scanning of Ga-68 DOTATATE is 40–90 min.
48. (a) True, tumors that do not bear somatostatin receptors will not visualize on the PET imaging.
49. (b) False, F-18 PSMA is not for imaging female patients as it is utilized to image patients with prostate cancer.
50. (a) The half-life of Ga-68 DOTATATE is 68 min.
51. (a) True, patients receiving F-18 PSMA for PET imaging are encouraged to increase hydration and frequency of urination.
52. (a) Short-acting analogs can be used up to 24 h before the tracer injection for the Ga-68 DOTATATE PET scan.
53. (b) Ga-68 DOTATATE is administered via intravenous injection.

36 Appendix 18: Answers to Chapter 18

1. (e) In addition to providing a map of tissue attenuation values, the CT serves to localize features seen on the SPECT or PET.
2. (b) X-rays produced by the CT are lower in energy than the photons emitted by radiopharmaceuticals, so attenuation coefficients derived from CTs must be scaled to the energy of the radiopharmaceutical. Because SPECT and PET images have poorer spatial resolution than CT, the CT must also be rebinned into large pixels or blurred to avoid artifacts at tissue boundaries.
3. (b) Anatomic coverage
4. (d) MRI scanners also have better soft tissue contrast than CT.
5. (b) You would modify the CT parameters in an effort to reduce the patient's radiation dose.
6. (f) The scan to define the body area to be scanned can be called either a scout or topogram.
7. (d) CT artifacts can be caused by the individual operating the scanner, the scanner itself, or the patient. They can also be caused by a variety of other factors such as the reconstruction algorithm and contrast medium.
8. (a) CT calibration must be checked daily, typically by scanning a water phantom, then (1) comparing the measured HUs to the permitted range of values, and (2) visually checking the images for artifacts. A tube warm-up may be necessary throughout the day depending on the scanner use.

E. Mantel et al., *Nuclear Medicine Technology*,
https://doi.org/10.1007/978-3-031-26720-8_36

9. (b) IV contrast may be administered but is not routinely administered to all patients having a SPECT/CT or PET/CT scan, especially if the CT is for attenuation correction only.
10. (b) Most commonly seen is the "banana" artifact, a curvilinear cold artifact just above the dome of the diaphragm/liver.
11. (a) and (b)
12. (b) and (c)
13. (b) The arrow is pointing to the outline of the patient's jaw in the PET image, and the mismatch between the jaw position in the PET compared to the CT indicates that the patient moved his or her head between modalities.

37 Appendix 19: Mock Examination Answers

Answers to mock exam

1. (a)	17. (c)	33. (b)	49. (c)	65. (c)	81. (e)
2. (b)	18. (b)	34. (c)	50. (d)	66. (c)	82. (d)
3. (b)	19. (b)	35. (b)	51. (a)	67. (d)	83. (c)
4. (b)	20. (b)	36. (d)	52. (c)	68. (d)	84. (d)
5. (d)	21. (b)	37. (b)	53. (a)	69. (c)	85. (c)
6. (b)	22. (b)	38. (b)	54. (b)	70. (a)	86. (a)
7. (c)	23. (b)	39. (b)	55. (d)	71. (c)	87. (d)
8. (b)	24. (a)	40. (a)	56. (d)	72. (c)	88. (b)
9. (d)	25. (d)	41. (b)	57. (c)	73. (c)	89. (a)
10. (c)	26. (b)	42. (d)	58. (a)	74. (a)	90. (a)
11. (c)	27. (e)	43. (a)	59. (e)	75. (d)	91. (a)
12. (b)	28. (b)	44. (b)	60. (b)	76. (d)	92. (c)
13. (b)	29. (e)	45. (a)	61. (b)	77. (a)	93. (d)
14. (e)	30. (d)	46. (b)	62. (b)	78. (b)	94. (a)
15. (c)	31. (b)	47. (b)	63. (b)	79. (d)	95. (c)
16. (b)	32. (a)	48. (e)	64. (b)	80. (d)	96. (b)

E. Mantel et al., *Nuclear Medicine Technology*,
https://doi.org/10.1007/978-3-031-26720-8_37